BIOLOGY IN HUMAN AFFAIRS

BIOLOGY
IN HUMAN AFFAIRS

by

WALTER V. BINGHAM   DONALD F. JONES

HUGH S. CUMMING   E. KENNERLY MARSHALL, JR.

EDWARD M. EAST   ELMER V. McCOLLUM

MORRIS FISHBEIN   HOWARD M. PARSHLEY

FRANK H. HANKINS   ARTHUR M. STIMSON

JOSEPH JASTROW   LEWIS M. TERMAN

Edited by
EDWARD M. EAST

Essay Index Reprint Series

BOOKS FOR LIBRARIES PRESS
FREEPORT, NEW YORK

First Published 1931
Reprinted 1972

Library of Congress Cataloging in Publication Data

East, Edward Murray, 1879-1938, ed.
    Biology in human affairs.

    (Essay index reprint series)
    Reprint of the 1931 ed.
    1. Biology--Addresses, essays, lectures. 2. Psy-
chology--Addresses, essays, lectures. 3. Medicine--
Addresses, essays, lectures. 4. Hygiene, Public--
Addresses, essays, lectures. I. Bingham, Walter Van
Dyke, 1880-1952. II. Title.
QH302.E3 1972   574   72-313
ISBN 0-8369-2790-7

PRINTED IN THE UNITED STATES OF AMERICA
BY
NEW WORLD BOOK MANUFACTURING CO., INC.
HALLANDALE, FLORIDA 33009

LIST OF CONTRIBUTORS

WALTER V. BINGHAM, PH.D., Sc.D. Director, Personnel Research Federation, New York. Formerly Professor of Psychology, Carnegie Institute of Technology and President The Psychological Corporation.

HUGH S. CUMMING, M.D., Sc.D. Surgeon-General, United States Public Health Service.

EDWARD M. EAST, PH.D., LL.D. Professor of Genetics, Harvard University.

MORRIS FISHBEIN, M.D. Associate Professor of Clinical Medicine, Rush Medical College. Editor, Journal of American Medical Association.

FRANK H. HANKINS, PH.D. Professor of Sociology, Smith College.

JOSEPH JASTROW, PH.D., LL.D. Lecturer, New School for Social Research. Formerly Professor of Psychology, University of Wisconsin.

DONALD F. JONES, Sc.D. Geneticist, Connecticut Agricultural Experiment Station. Editor, *Genetics*.

E. KENNERLY MARSHALL, JR., M.D., PH.D. Professor of Physiology, Johns Hopkins University.

ELMER V. McCOLLUM, PH.D., Sc.D. Professor of Biochemistry, Johns Hopkins University.

HOWARD M. PARSHLEY, Sc.D. Professor of Zoology, Smith College.

ARTHUR M. STIMSON, M.D. Assistant Surgeon-General, United States Public Health Service.

LEWIS M. TERMAN, PH.D., LL.D. Professor of Psychology, Stanford University.

[v]

PREFACE

THE emotional and intellectual unrest of our day is too often treated as a novel and disquieting phenomenon indicating the imminent collapse of the whole social structure. The situation exists, it is true; yet it is hardly a cause for alarm, except as it stimulates the quacking of so many anserine saviors. All peoples, in all ages, have been in similar states of flux, for the simple reason that there have always been new ideas demanding attention; and where there are new ideas, there are reactions against them. As a means of retarding the entry of unusual notions, the human cranium is a well-nigh perfect mechanism. Were we more analytically inclined, we should see in such manifestations merely an evidence of progress. The old order passes, and we are reluctant to have it go; our agitation, therefore, marks at once our feeble adaptability and the degree of mental adjustment imposed by the occasion. It may be true that man now confronts a situation requiring a greater change in attitude of mind than any he has faced before, since he is asked to abandon his most primitive, and therefore his dearest, superstitions; but then, he is better prepared through steady practice.

Since the time when Leonardo da Vinci· set the true pattern for profiting by experience, some four hundred years ago, the intellectual horizon of mankind has constantly broadened. One by one, traditional illusions have been replaced by demonstrated truth. In turn, mathematical, physical, and chemical concepts have guided human thought into new channels, in spite of all checks and hindrances. And now a fourth group of scholars asks that biological ideas be received on terms of equality with those of the other sciences. The biologist believes that he has much to offer, both to man the intelligent being and to man the animal, subject to hunger and disease. He is convinced that even the most intimate problems of existence, those problems which concern man's mental and emotional life, are susceptible of a helpful objective analysis. At the same time, he realizes just how difficult it is for most people to accept the thesis that the complex attributes and the many-sided activities of the human being can be studied as carefully, as precisely, and as impartially as those of any other being about which we wish to learn the truth. It requires some knowledge of what has been accomplished, some familiarity with the methods by which the work is done; and it demands an intellectual independence above the ordinary.

As a contribution toward a better understanding of the ideals, the methods, and the results of biology, this book is presented. It contains a series of relatively non-technical discussions of the present conditions of several of the more important subdivisions of the subject, emphasizing recent advances and expressing a few conservative predictions as to probable future trends. Those phases of biology which have been—or may be—of especial service to mankind have been stressed, with the hope of enlisting the interest of the general reader and of giving him a clearer and more accurate idea of the problems involved and of the progress in solving them. In one sense, therefore, the authors have endeavored to give a wider publicity to their wares. But there has been no insulting attempt to tell nature stories of the type which appeals to immature minds. It is a serious book intended for people who find satisfaction in the disentanglement of the mysteries of the universe in which they live.

The subject matter falls naturally into three parts. The first part is general in character. It contains reflections upon the philosophy of biology and upon the validity of its conclusions. Here the authors have also tried to show how the field of biology is ever broadening through the inclusion of problems which were formerly thought to be beyond the reach of scientific procedure. The second part is concerned with the mind. It contains a skeletonized account of the general development of psychology, and essays on the application of psychological methods to education and to industry. The third part treats of the body. It is designed to show, in some measure, the progress in dealing with man as an animal. It portrays some of the advances in genetics, in medicine, in the work of the public health officer, in physiology, in biochemistry, and in general zoology put to the service. In addition, there is an extended discussion of the means by which man increases his food supply and thus makes it possible to feed some twenty million additional people with the advent of each new season.

E. M. E.

Boston, Massachusetts,
February, 1931.

CONTENTS

Chapter I

BIOLOGY AND HUMAN PROBLEMS

by EDWARD M. EAST

DURING the past century and a quarter science has remodeled man's universe. In this period, so short that many of us have talked with those who saw its onset, more significant discoveries have been made than in all previous ages combined. One cannot picture vividly the vast changes induced by probing the secrets of atom, molecule, and cell. Comfortable dwellings, luxurious journeys, instantaneous communication, sanitary conditions, longer and healthier spans of life, economic and intellectual freedom, are accepted casually, without emotion, without thanksgiving. We think of them as part of our natural environment—like clouds or sunshine or green trees. But conditions were quite different at the beginning of the nineteenth century, so different that an evening with the most impressively written history of the time fails to carry us back and to identify us with the situation that confronted our great-grandparents. Perhaps a visit to central Africa might do it, but few of us have such an opportunity.

Unfortunately science has not recast man's thinking to anything like the same degree. The average mind is still fettered to myths conceived in distant eras of unreason. The proportion of persons guided by rational ideas has certainly increased in recent years; but an intellectual lag obtains which shows in even the best of us. The trouble is that the whole world profits from all new discoveries, yet only the exceptional few care about the philosophy of thought which makes them possible.

[1]

Intelligence tests carried on in the more enlightened countries during the past two decades have shown that fully fifty per cent of the people involved lack the capacity to follow even elementary scientific reasoning. This fact is often accepted as evidence that the scientific outlook can appeal to only a small fraction of the population. No conclusion could be more superficial. The average mind, the mean mind, as the statistical wag puts it, absorbs teaching to the limit of its power. It is conditioned by its experience, as all minds are conditioned by experience. But where the superior mind is compelled—also by training—to sift its varied grist of lore, holding fast that which is good, the untrained or inferior mind takes in more indiscriminately whatever is offered of digestible quality. The reason the brains of most people are stuffed with so much medieval nonsense, therefore, is that this is the intellectual fodder with which they have been supplied. Clearly, then, if progress in the world of ideas is to keep pace with material progress, these thob-fancies must be replaced by ascertained truth.

Can it be done? Well, why not? Have we no leaders of thought? Can we make no changes in our educational system? It *is* very difficult to alter the *mores* of a people, as Sumner has shown in that noble treatise "Folkways"; yet it is not impossible. The soviet rulers are doing it rapidly and thoroughly. It is true that here one set of foolish ideas is being replaced with another; but this fact is of little consequence. The important thing is the demonstration that success is possible in such an undertaking. My personal belief and, I think, the universal belief of all true men of science, is that humanity can aspire to nothing more lofty or more satisfactory than a philosophy based solely on demonstrated truth. I should like to see a world in which people endeavor to deal only with facts, all the facts, and nothing but the facts; where truth is approached without prejudice, and the course of action to which it leads is faced without evasion. I believe that it would be a more comforting as well as a more comfortable world.

Such creedal statements immediately stimulate emotional reactions in people whose minds have fixations on the mystical. Before they burst forth in criticism or denunciation, I have always wished that they would answer one question: Is it or is it not a worthy objective to prefer truth to falsehood? But they are too cunning to be caught in such a simple trap. Instead, they suggest that the above doctrine is "the substance of things hoped for," and therefore a faith rather than a belief; they say that they also are seekers after truth, but that their brand of verity includes subjective truth, intuitive truth; and they maintain that numerous able scientists hold similar opinions.

H. L. Mencken, with his usual forcefulness, has shown the absurdity of the mystics who thunder against the beliefs of men of science as being indistinguishable from articles of faith. "Belief is faith in something that is known," says he, "faith is belief in something that is not known." The distinction is good; and by it the faith of the scientific man in objective truth, established by impartial evidence and demonstrable to all who have been taught the use of reason, becomes a belief. The basis for it is the fact that the whole course of human progress has been due wholly and solely to the use of scientific methodology. There is no more justification for faith in intuition than for faith in witches. Intuition is merely a delusive word for something that wishful thinkers hope for, a phantom to enable them to vindicate their own irrationality. There are instincts, it is true, but experimental physiology has shown that instincts are drives which are tropistic in nature and which have little connection with the higher cerebral centers. They should not be confounded with intuitions. For example, certain larvae, which hatch in

[3]

loam, make their way immediately to the tree tops. But they do not take this journey because their brains tell them, in some mysterious intuitive way, that food is there. They go because their optical outfit is so constituted that they are driven toward the light. Desensitize their eyes, and they starve to death, though a well-filled larder is near at hand.

As to there being true scientists whose tenets include belief in the possibilities of attaining knowledge through some mysterious insight which differs from all ordinary percepts and their rational consideration, I simply deny the allegation. It is a mistake to assume that when Millikan, K. T. Compton, and Pupin issue preachments in terms of theology and metaphysics, they are speaking as scientists. They are merely demonstrating how difficult it is to divest one's mind completely of the whams and whimseys learned in early childhood. These men are competent physicists who have done admirable work in their own bailiwick by using the objective methods of science. This procedure has worked well in the solution of the problems in which they have been interested. Through it they have been successful. By its means they have attained fame. Do they give honor where honor is due? Oddly enough, they do not. In their leisure hours they endeavor to show that the perplexing questions which concern man most intimately— problems of emotion, of conduct, of thought— are unassailable by these time-tried methods. They speak vaguely of man's higher nature, something above the rational, which may attain to truth by mysterious inspiration. And they are speaking of man, you understand, of that animal of the family Hominidae, whose corporeal make-up is better known than that of any other organism; whose functional activities have been the subject of experimentation for a century and a half; whose hopes, whose desires, whose emotions, whose mental peculiarities are objective problems about

[4]

which the psychologist knows almost as much as the physicist knows about the atom; and all through the application of scientific methodology. In other words, these men who have become distinguished by using scientific methods in a given field, turn to a department in which they have no accurate knowledge and maintain seriously that here the primitive folklore of the race, unsupported by the type of evidence they would demand in their own researches, is preferable to the inductive conclusions reached by scholars working objectively on the subjects involved. In doing so they merely show that a man may do excellent work in science without being a scientist at heart.

These statements— or any others I may make in an insistence that the scientific attitude of mind is the only position worthy of the species, if it is to be known as *Homo sapiens*— will convince no adult who has reached different conclusions previously. If a man of mature mind has become assured that science is materialistic and interferes with the spiritual life, it is useless to try to change his views. But this fact by no means releases the scientist from the duty of propagating his beliefs. The impact of his arguments upon younger minds, still incompletely hypnotized by the imagery of metaphysics, will eventually yield results. These more elastic individuals will come to realize that man is to be distinguished from the other animals on the footstool chiefly by his greater intelligence, and that only by the exercise of this intelligence can he have a higher life. They will see that the mystical pother over man's apartness from the rest of creation is merely a formal expression of the hopes and desires arising from his superbly egocentric imagination. And as such disciples grow in numbers, so will the momentum of intellectual progress increase in strength.

One has the feeling that science is advancing and bigotry retreating at a more rapid pace to-day than was the case a short while ago. Anti-evolution resolutions are still presented to the Solons of the various states, and Billy Sunday is still greeted with applause; but science is in distinctly better odor than of old. Even the newspapers, as always the rear guard of public opinion, speak of the current era as the Age of Science. The present generation has almost forgotten the battles by which the cohorts of mysticism formerly tried to prevent the acceptance of every discovery in biology, geography, astronomy, geology, philology, anthropology, and medicine, which appeared to run contrary to dogmas of primitive faith. I invite its members to read the history of the campaign in the two ample volumes compiled by Andrew D. White.[1] The struggle is described there in all its gory detail.

I am not convinced, however, that science has sustained these attacks because truth has an inherent survival value. Science was retained on sufferance for the same reason that a fine old peach tree is retained—people enjoyed its fruits. The scientist need have no fear, therefore, that his freedom of action will again be curtailed as it was in earlier days: the products of his labors are too eagerly welcomed. Nor am I convinced that the spirit of science, the scientific attitude of mind, is any more acceptable to the defenders of dogma than it ever was. Yet open warfare against science has been abandoned by all except the most ignorant minority of the clerical party; the leaders have found it advantageous to take a more dignified position and to leave acrimonious debate to others.

The current fashion is to put the case against science into the hands of the philosophers, who habitually issue verdicts with as much solemn formality as if they were presiding at the Last Judgment. Statements constantly appear to the effect that Bergson or Boutroux or Whitehead or

[1] WHITE, ANDREW D., "A History of the Warfare of Science with Theology in Christendom," Appleton, New York, 1896.

[6]

some other metaphysician has signed a death certificate for Scientific Naturalism or Neo-mechanism, as the philosophers so kindly label the collective efforts of the humble workers who have solved some small portion of the problems confronting mankind. These announcements, like similar announcements of the past, are premature. The patients, despite their treatment by these Borgian physicians, remain sufficiently robust to continue recording the collapse of metaphysical hypotheses.

If one scans the history of the subject, he can readily sympathize with the reaction against science exhibited by the modern mystic in cap and gown who still answers to the name "philosopher." Formerly philosophy took all constructive thought as its province. It attracted great minds to the task of analyzing and synthesizing knowledge; and these illustrious scholars left a legacy for which we, the recipients of their bounty, may well be thankful. Then came the decay of near-eastern civilization, and the Dark Ages. With the revival of learning, philosophers became divided into two camps. One group heeded the demand of Aristotle "to trust more to observation than to speculation, and to the latter no further than it agrees with the phenomena"; they dropped the ancient title and adopted that of science. The second group retained the noble old patronym and, disdaining all vulgar use of hand and eye, settled themselves to subjective rumination on ultimate causes. To-day the impotent residua of this clan wander through the ruins of their lost estates bewailing that nothing remains but the ghostly outlines of metaphysics, the rusty toys of epistemology, and the vestiges of an ethics which has lost all influence and prestige.

What other fate could the philosopher expect? As I have said in another essay, he yearned to be the gentleman, not the mere private secretary to Nature. He wished to evade the drudgery of the patient collector of facts, of the experimenter, of the calculator—whose wages, though sure, were small. He was enticed by the visions of wealth that speculation always promises so impudently. Consequently, he has gone bankrupt, like many another who has hoped to become rich quickly.

The philosopher, probing the universe for final causes, resents the success of his kinsman who has vulgarly engaged in trade. Under the influence of the delusion that he is grand arbiter by divine right on all intellectual problems, he tells the latter that his methods are limited in application, restricted in validity, and thus hopeless for the purpose of detecting absolute truth. Such criticisms of science cannot be refuted. They are true. But they do not have quite the consequences assumed by even their most logical proponents, and they certainly give no support to the doctrine of self-evident truth. They fail lamentably, therefore, for they are put forward chiefly as props to intuitionalism.

Let us look a little further into these questions. By making certain assumptions, any good metaphysician can prove that the earth does not exist; by making somewhat different assumptions, he can prove that it does exist, though only because man is present to perceive its existence. Now the scientist doesn't care the slightest whether the earth exists or not. He wants to know whether it is flat or round; he wants to know its constituent parts and contents; and by assuming that it does exist, he is able to do these things with some degree of satisfaction. The illustration is silly, despite the fact that solemn volumes have been written on the points involved; but it does serve to show that all formal systems of logic have the same faults, in that they must all begin with assumptions. The scientist realizes that his work has the faults of all systems in just this respect. For this reason he has long since ceased to worry about absolute truth. In his opinion absolute truth is a metaphysical term with no more meaning than abracadabra. He has come to realize that the only form of truth worth considering is what one might call man-truth. This form of verity is based upon man's belief in the time-tried doctrine of causation; it is tested by the five senses; it is sifted by experience. It is a simple scheme, that of acquiring knowledge by describing phenomena in terms with which we are familiar and of determining the relation between them; but it is the only one which has been found to be continuously and invariably serviceable.

Science makes, or endeavors to make, full use of the human faculties; and, obviously, they are the only tools at hand. But man is an imperfect creature; hence his gropings have their limitations. Unfortunately, as soon as one admits that scientific truth is relative and that the method has its limits, these admissions are stretched and distorted in a wholly illogical way. It is assumed that all findings of science have the same degree of probability—or improbability. Only too frequently science as a whole is held up to ridicule because of the discarded hypotheses of physics, chemistry, and biology. To scoff at these dead soldiers is ill mannered, for many of them did good service in the scientific advance; but to try to persuade the uncritical to believe that all scientific discoveries are similarly short-lived is criminal. For one insufficiently established theory there are thousands of facts which will never be questioned. They will stand as written, once and for all.

It is also a popular pastime, particularly among those most unfamiliar with the situation, to stake out the limits to which science admits. The sign "No trespassing here" is as familiar in the history of progressive thought as it is along the country roadside. In the past it has been placed over the estates of chemistry and physics; more recently it has been shifted surreptitiously to the domains of biology. Scientists have been told to restrain their curiosity regarding heredity and intelligence and consciousness.

[9]

Such subjects were divine mysteries. But members of the Promethean guild have never been impressed by signs. The results of their boldness are the sciences of genetics and psychology. The plain truth is that the entire universe of human experience is attackable by the scientific method; and when the last human succumbs to the freezing temperatures of this cooling sphere, he will leave solvable problems still unformulated.

The same timid soul who wants to preserve the Olympian mysteries inviolate is also disturbed by what he calls scientific dogmatism. The scientist is dogmatic; it must be admitted; but he is not dogmatic in the sense implied by his critics. He holds no inflexible confidence in plausible conclusions, insecurely founded, just because they bear the label of his trade union. If a new theory appears which is better substantiated than the one he holds, he will give the latter up at once. His only determined position is concerned with the validity of the scientific method, with the spirit of science. He holds that the one means of solving human problems is further advances in science. He takes this position for the simple reason that, in all its cycles of experience, the race has found no other method that really works. It is a substantiated belief and not a hope. If this is dogmatism, make the most of it.

We have recently had an episode of such criticisms of science that was rather interesting while it lasted. I refer to the rise and fall of the Humanist cult led by Irving Babbitt. I mention it only because the leaders of the movement have been literary men and have therefore been able to advertize their doctrine widely, and because the slovenly formulation of the ideas involved has led many readers to think they see in them a depth and value altogether undeserved. Humanism is the name of an undertaking begun about the year 1300 in which the goal was the discovery and promulgation of the good life on this lowly planet by the use of

[10]

human faculties. With such announced intentions one might assume, it seems to me, that scientific work should have been promoted. Galileo and Kepler and Harvey should be the pride of the order, since ability to reason is certainly the distinctively human trait. But no; not even such master deriders of folly as Boccaccio and Rabelais are admitted to fellowship. The true humanist, as nearly as I can gather from the unintelligible though voluminous definitions of the modern exponent, is not he who bestows new knowledge upon his fellow men, but rather he who believes that progress ended with the decline of the classic Greek tradition. The announced tenets of humanism are a sense of decorum and a "will to refrain"; and each humanist naturally writes a book to explain what these precepts mean. The resulting volumes achieve harmony on only one point, and this makes the contents far from dull: there is no need for restraint when attacking error.

Out of the modern humanist doctrine there somehow emerges an acute dislike for science. Professor Babbitt says that science is well enough in its place, but demands for the humanist the right to determine its place. The difficulty with science, according to L. T. More, who has received some training in the older type of physics, is that instead of confining itself to the simple problems of the atom and the molecule, it is also studying human attributes and emotions. It does not appear that he requires the closing of all medical schools, but he does urge fetters for all psychologists.

What has led to such queer and loose-jointed thinking? As Walter Lippmann has so ably shown, the immediate trouble with these gentlemen is that they have adopted as dogma an assumption long since disproved; namely, that there are determinative laws for things—and presumably for the lower organisms—but none for man. Having asserted this dogma, they hold it to be a demonstrated fact.

[11]

This interpretation is accurate, but it does not probe deeply. The stimulus which motivates the Babbitt group is an inferiority fixation. I dislike using the term because of its appropriation by the cultist Freudians; but it has a validity in the interpretation of certain types of behavior that is not to be denied. The inferiority complex is at the bottom of most tirades against science. In the particular case under discussion, the thing is acute. Nearly all present-day humanists are literary critics. Now Chesterton, who is sound enough when he sticks to professional items, says that learning to write is so difficult that "it prevents general 'education' in the customary sense of the term." The writer, therefore, is a privileged man, since "he is permitted to have no exact knowledge of anything in life—except his craft, and not always that." The literary critic is naturally in a worse case than the creative writer in this respect. Moreover, the critic gets in the habit of speaking with such cocksureness about literary artistry that it is only natural for him to react more violently still against creators of knowledge whose ways are a closed book.

Deeper still there is a reaction common to nearly all of us. It is the protest against the lowly place in the universe which the still, small voice of reason gives to man. Astronomers have shown the probable origin of the earth; geologists have divulged its history. Chemists have manufactured compounds which formerly were thought to be characteristic products of organic processes. Biologists have linked man with the humbler animals in structure, function, and behavior, thus demonstrating by overwhelming evidence the doctrine of evolution. Geneticists, psychologists, and physiologists have combined their evidence and seriously assert that man's individuality is the necessary product of his hereditary constitution plus physical environment and experience. It is perhaps to be expected that we who have created gods in our image should rebel

[12]

at such tenets, and should declare that we will accept no conclusions which rob us of spiritual and moral values and leave us like so many automata, bereft of faith and hope, tossing helplessly amid other combinations of electrons in a mechanical universe ruled by chance. These things, we say with Jurgen, are lunacies of realists which we do not choose to believe. "For how shall I believe," says Cabell's hero, "that all men who have ever lived or will ever live hereafter—that even I, am of no importance?"

Man would rather feel that his ego holds high value in the macrocosm. He aspires to a dominant position in this life and hopes for immortality hereafter. Sorely troubled with the dawning suspicion that his destiny may be quite different from what he has promised himself, he attributes his disillusionment to science. No, I am wrong; he makes biology the scapegoat to-day; he no longer cares about the generalizations of the physical sciences, having quite forgotten his plaints against their destroying discoveries a century ago.

There is no justice in such accusations against organic science. The more we know about the human race, the more absurd its pretensions appear to be. One cannot dodge that issue. Yet science has nothing to say against man's values here and now; it has nothing to say against values after death. It merely asks that intelligence, morals, and conduct be studied along with structure, function, and environment, as a prerequisite for dealing more rationally with mental, moral, and physical adjustment. I do not know what proportion of the stimulus to scientific work is curiosity, what proportion is will to power, and what proportion is humanitarian desire; but I do know that the results make the world a better place in which to live.

In early societies man cowered in fear and awe before the wonders of nature. He was afflicted with illnesses of body and of mind, oppressed by forces of which he was ignorant; and he bowed in supplication before the powers behind them. With the lapse of time and the influx of knowledge, he became more courageous and more arrogant, and these primitive beliefs changed. Yet man finds himself still perplexed, still oppressed, still pitting his paltry strength of mind and arm against stern reality. His refuge from the unhappiness with which he is thus confronted is to construct subjective universes which are kinder to him than the one in which he is actually placed. This is the strategem of the theologian, the mystic, the philosopher, the humanist, the artist.

But a second course is open. We can study mundane phenomena impartially. We can find out, not what they really are or what they really mean, but what they are to us and what they signify to us. This is the method of science. Through it mankind has already been brought from savagery to civilization, from harassing struggle to comparative comfort. Through it the race may approach peace and serenity of soul. It is certainly no wild delusion to assume that whatever happiness lies within man's grasp will come when the machine has banished the fear of want, when each individual can face life with his mind properly trained and his body kept in constant trim. Perhaps this goal will not be reached, but there is every reason to believe that it will be approached. Sufficient knowledge is now extant to satisfy all economic needs were it to be applied in its entirety. We are not ready, at the moment, to prescribe for all ills of the body or to make all the necessary mental adjustments; but biology is young. Surely the time will come when the mischievous *conditioning* of the mind through false teaching, which now causes so much unrest, will be a thing of the past.

Objection will naturally be brought against this program. The flesh is weak, the critics will say, and the mind is weaker. Our adult infants must always endeavor to evade reality by indulging in happy daydreams. Of course they must! No one would think of robbing them of such a refuge. They will always have avenues of escape into the unreal; for who would banish literature, drama, art, and music from the world? What the scientist objects to, and what proper instruction can prevent in persons of normal mentality, is the tendency to believe that these make-believe worlds are tangible realities. "I believe that an artist, fashioning his imaginary worlds out of his own agony and ecstasy, is a benefactor to all of us," remarks Mencken, "but that the worst error we can commit is to mistake his imaginary worlds for the real one."

I have endeavored to describe the methods and ideals of science, to show its unity of purpose, to demonstrate its efficacy in the solution of the more intensely human problems. My aim was to emphasize the fact that biological questions must be approached in the same spirit as other questions, even when human behavior is the issue; for man can be studied quite as objectively, quite as satisfactorily, as any other phenomenon of nature. The physical sciences have given us no mean conception of the affairs of the material world; biology must give us the power to understand ourselves and to control our destinies. In the following pages, my colleagues and I will attempt to offer the reader a glimpse of the present status of certain subdivisions of the sciences dealing with life. Obviously the accounts presented will be incomplete, owing to spatial limitations, and hence will give no adequate idea of biological progress. Ineffectual as they may be, however, they will show that the biologist has gone far. As an augury of greater things to come, therefore, I wish to devote a few pages to the history of biology, in order to draw attention to the extraordinary rapidity with which biological discoveries have been made in recent years. Biology is still an infant whose precocious efforts should lead us to expect great performances during youth and manhood. I shall not apologize to the professional scholar into whose hands the book may fall, for recalling familiar facts, since such a reconnaissance, emphasizing as it does the merit of modern work, may serve to bolster up his faltering spirit on some Blue Monday.

Apart from what is found in the works of Aristotle, the biology of the ancients was medicine. The medicine, for a thousand years of our era, was that of the second century Greek physician Galen, a healer who was ignorant of even the correct number and placement of the bones of the human body. The Byzantines made some surgical advances, and the Arabs developed the use of drugs in the treatment of disease; yet Galen remained the only real medical authority of Christendom during the whole of this black millenium. The frightful superstition of the times smothered all originality of thought; hence, the infrequent and relatively unimportant discoveries recorded led to no philosophical changes. The greater part of Europe, in fact, was in such a state intellectually that, by the beginning of the thirteenth century, the whole great legacy of learning left by the Greeks was becoming a distorted tradition. To restore classical authority was an objective of humanism; and at least one physician of this period should be acceptable as a colleague to the Humanists of to-day, for Arnald of Villanova dedicated his life to divesting the current leechcraft of its accumulation of worse errors and to re-establishing in pristine purity the teachings of Hippocrates and Galen. But there was a better way than that of Humanism. Reference to authority, whether of Greece or of Arabia, could not lead onward. The course of progress is inquiry, and all through the next three centuries a few doughty souls took this course. They should command our respect, even though they made no outstanding discoveries; for they sought for truth where truth can be

[16]

found, ofttimes at considerable cost. Peter of Abano, one of the first great lights of Padua, was persecuted for expressing the belief that diseases had *natural* rather than *supernatural* causes, and avoided the fagots of the church only by dying opportunely.

Obviously the organic sciences had no structure at the beginning of the sixteenth century. There existed only the Greek foundation, frayed and weather-beaten. The New World, therefore, is no newer than biology. It is no more than justice to say that biology, in the true sense, began when Vesalius the Fleming, just turned twenty-three, was given the chair of surgery and anatomy at the University of Padua in 1537. Vesalius cast aside authority and, trusting solely to his own observations, labored to such purpose that in five years he was able to publish the monumental folio "*De Humani Corporis Fabrica*," illustrated with the beautiful drawings of that gifted student of Titian, John de Calcar. His work received the vigorous opposition usually accorded to innovators. He had found no incorruptible resurrection bone which theologians had held must exist; he had discovered that man has the same number of ribs on each side of the body, to the great horror of the students of Genesis; and he had caught the ancient master, Galen, in numerous errors. But the efforts of the opposition had little effect. Vesalius, together with his lesser contemporaries, Eustachius and Fallopius, had established anatomy as an observational science; and shortly after, with the aid of the newly discovered microscope, it was carried to great heights by Malpighi in Italy, by Leeuwenhoek, Swammerdam, and Lyonet in Holland, by Réaumur in France, and by Hooke and Grew in England. Probably no greater morphological work has ever been completed than the entomological dissections made by the brilliant Italian and his distinguished co-workers from the Low Countries.

[17]

In the meantime biology had set another milestone in its course. William Harvey (1578–1667) had demonstrated the circulation of the blood, and thus had laid a foundation for the quantitative study of physiological processes. Without the aid of the microscope Harvey was unable to trace the capillaries and to interpret their function, this discovery falling to the lot of Malpighi six years before Harvey's death. But Harvey did something better. He used his reasoning power. By observing carefully the action of the heart and the difference in structure of the veins and the arteries, he was able to show definitely that the quantity of blood pumped into the aorta in a given time makes its return to the heart necessary, since in less than half an hour a quantity of blood greater than that contained in the whole body passes into the great artery. The consequences of this discovery were epoch-making, in that it was fundamental to a true idea of respiration and hence to the proper conception of the body chemistry. But it was more than this. It showed that biology could pass beyond the mere classificatory stage and begin to compare relationships between observed phenomena.

During the latter part of the seventeenth century, the whole of the eighteenth century, and the first quarter of the nineteenth century, biology was by no means stationary. An enormous quantity of carefully collated observations were given to the public. Life histories of thousands of organisms were completed. The organic world was enlarged by the discovery of protozoa and bacteria. Botany flourished. Linnaeus put the classification of animals and plants on a sounder and more satisfactory basis. Lamarck endeavored to demonstrate organic evolution. Nevertheless, biology had not learned to walk, or to run, as it does at the present day; it still crept. Its progress was so slow that if one takes the total biological knowledge of to-day—weighted in proportion to its importance—as one hundred,

[18]

ninety-nine parts of this knowledge have been obtained during the past century, the full span of a single human life. This fact makes the future seem quite rosy.

In 1828 a new field was entered. Wöhler made urea in his laboratory. For the first time a product, until that day known only as the waste of animal activity, was produced by artifice. Henceforth *all* organic activity was interpretable in the terms of chemistry. To-day the whole sequence of chemical changes from the intake of raw food to the resultant bone and muscle is pretty well mapped out. The way in which the food breaks down during digestion, the manner in which the blood acts as carrier of the resultant constituents, and the method by which even the synthesis of proteids occurs, can all be traced with a considerable degree of accuracy. And from the knowledge gained through studying normal metabolism, our biochemists have learned much concerning the abnormal. Medicine has profited. Pathologists already utilize chemical tests to discover whether organs are functioning properly; and the day is not far away when much organic deficiency— whether hereditary or the result of disease—can be made good by the compounds of the commercial chemist. Insulin is the herald of these good tidings.

In the third decade of the nineteenth century, biology reached out again. Schleiden and Schwann demonstrated that cells are the structural units of every organism, whether animal or plant. This discovery, denounced in no uncertain terms as an impossible theory for several years, revised our whole concept of living things. It was, perhaps, more important as a philosophical innovation than as the opening investigation of the field of research now termed cytology; nevertheless the study of the cell has yielded concrete results of extraordinary value, considering the limited period during which it has been carried on. Our leading investigator in this department of knowledge,

Edmund B. Wilson, has happily had the individual experience of seeing the establishment of the theory that all cells come from pre-existing cells; of seeing the proper description of cell division and of fertilization; the identification of the cell nucleus as the most important cell organ; the discovery that the chromosomes, with their peculiarly exact methods of division, are the material basis of heredity; and finally, of seeing the determination of the architecture of the germ plasm and of the mechanism by which the units of heredity, the genes, are transmitted to the succeeding generation.

Physiology, as I have said, had its birth in the researches of Harvey; but for two centuries and a half it led a rather moribund existence. It was restored to vigor by the genius of Johannes Müller (1801–1858), of whom Verworn said: "He is one of those monumental figures that the history of each science brings forth but once." Müller was not only a great investigator, he was a most stimulating teacher. Du Bois-Reymond and Helmholz are two of the distinguished scholars whom he led into scientific work. Müller's contributions to physiology were prodigious, yet he is revered to-day less for the specific results of his researches than for his innovations in physiological methodology. He made controlled experiments, he studied animal behavior, he introduced quantitative measurements, he interpreted results both in the terms of physics, chemistry, and mathematics, and in the terms of psychology. In fact, it was from Müller's laboratory that there came the first recognition of the close association between the operations of the mind and the functioning of the central nervous system.

Physiology, in the seventy-two years since Müller's death, has probably made the greatest progress of any of the biological subdivisions. The introduction of quantitative methods permitted the statistical study of sensation,

BIOLOGY AND HUMAN PROBLEMS

of tropistic response, and even of more complex types of behavior. It stimulated the more precise investigation of all sorts of bodily functions. It gave us a more accurate conception of the body as machine, and showed us how to accelerate or retard the wheels of the various parts. It showed us, moreover, the interdependence of these different parts and demonstrated the unity of the organism as a whole. It made possible such apparent miracles as the development of the egg without fertilization. It budded off colloidal chemistry and biophysics. It was the starting point of immunology and serology, those Siamese twins of science that have revolutionized medicine. Most important of all, because of the effect such demonstrations have on belief in the universal applicability of the scientific method, it proved the influence of the secretions of the body upon behavior.

One might dwell upon the rise of comparative anatomy, and histology, and embryology, and paleontology, and anthropology, upon the development of more natural systems of classification, upon life histories, upon geographical distribution. These subjects certainly are important. The knowledge such departments give us of the world in which we live is sound. It delivers a history whose accuracy is more to be depended upon than those social chronicles which come to us under that name. One could justify such studies even on economic grounds. But these sub-sciences—sciences of the classical type, as Ostwald calls them—have had little influence, by themselves, in changing man's intellectual habits. But this is not to their discredit. Collectively they furnished the evidence by which Charles Darwin proved the fact that Organic Evolution had occurred, a generalization which has done more to overthrow bigotry and install rationalism than any other. And when I use the words *proved* and *fact*, I use them intentionally. Only those unfamiliar with the evi-

[21]

dence have the impudence to claim that the Evolution Concept is an undemonstrated hypothesis.

Darwin showed how the chaotic series of facts relating to the geological succession of organisms, to geographical distribution of forms, to morphological similarities, to persistence of vestigial organs, to the recurrence of embryological phases in such distinct groups as mammals, birds, and reptiles, to change of form under domestication, were each and all, individually and collectively, brought into intelligible order *only* through the idea of evolution. Since these biological facts number hundreds of thousands, since they include the entire compendium of biological knowledge, in truth, I venture to call Darwin's generalization the greatest effort of the human mind. It is true that we do not know all that we should like to know about the way in which evolution took place, but that is another story; of the phenomenon itself we are assured.

The evolution concept stimulated every line of biological work. Through its guidance thousands of problems—upon which numerous observations had been made, but where the meaning of these observations was obscure—were problems no longer. Their solution was at hand. But these results, despite their value, were not the most significant effects. The impact of the idea upon the human mind is what has really counted. When the notion of the genetic unity of living things was once completely grasped, it became apparent that the temptation to set apart certain human problems as definitely beyond the reach of the inquiring mind was simply a manifestation of the primitive interdiction of taboo. Measured thus, the value of Darwin's work became incalculable. Single-handed he had rent the veil of mysticism which had impeded human progress from the beginning of time. No longer was it possible to maintain a holy of holies into which the scientist was not permitted to enter. It must be admitted that after sixty

[22]

years there is still a battle line defending the hallowed ground; but it is sadly harassed, and its morale is low. In places an entry has been made, and the invaders have consolidated their gains. The results are seen in sociology, genetics, and psychology. The historical researches of the sociologists—showing, as they have, that moral ideas are matters of custom—have opened up the way to a more acceptable framework of society based on reason and knowledge. The geneticists have solved the mechanical problem of heredity, and have gained a clear idea of the factors involved in somatic development. Thus a whole category of problems has been removed from the realm of the mysterious, taken to mundane precincts, and arranged ready for systematic attack. The psychologists have made sufficient progress to demonstrate that the activities of the mind have a physical basis. They have shown how easily sensation, perception, and even intelligence may be measured. They are fast coming to the point where they can tell us not only how we learn and when we learn, but also what we do with our learning and what our learning does with us. It is not too much to say that, within another generation or so, the most serious problem the human race can ever face, mental adjustment to reality, will have had at least a formal solution.

If this brief and ragged account of the efforts of biologists brings to the imagination no vision of a world less encompassed with sorrow, the fault is mine. Let us assume that the task had been accomplished to better purpose, for there has been progress, admirable and exhilarating progress. What then? Is it to be assumed that science can do away with all of the trouble of the human race and can transform this vale of tears into the best of all possible worlds? No! Such an eventuality is not to be expected, and if it were a possibility, the result would probably be quite unsatisfactory. Those havens of everlasting joy which our fancies have conceived as the last word in fulfilled desire have always seemed to me to be a little tiresome. There is in them no place for diversity—and variety is a very acceptable thing. The vision of society that advance in science, properly applied, raises in one's mind is a distinctly human society, after all; for the dissimilarity of individual endowments can never be eliminated, the drives of the basic instincts can never be wholly suppressed or sublimated, and man must get along as best he can with an imperfectly designed set of organs which will often function badly.

As I see it, the possibilities of Scientific Humanism, which must not be confused with Literary Humanism, run somewhat as follows:

Under a just and humane government the machine can abolish poverty. Indeed it can furnish creature comforts amounting to luxury, although with the inculcation of healthy ideals, inordinate demands for luxury may be expected to diminish. Birth control can lay the Malthusian spectre of overpopulation and keep the census figures at somewhere near the optimum for effective effort. Genetic information, sanely directed, can lessen the proportion of the mentally and physically deficient, and can raise the average intelligence materially. Medicine and surgery can increase the average expectation of life at birth to sixty or seventy years, and this span can be made relatively free from disease.

Murmurs of criticism will break out at this point, I dare say. "This is all very nice, but the chief object of such a program is to make man a sleek, healthy animal. Vice and corruption will still flourish." Not so fast! In a scientific world, the term *vice* will not be used. It will be replaced by the phrase *asocial conduct*. And asocial conduct will be studied and understood, prevented or forgiven. Moreover, behavior has a physical basis, though the immediate stimulus may be due to other factors. A con-

[24]

siderable proportion of crime is caused, primarily, by defects in the central nervous system. Even slight lapses from ideal behavior may be traced to physical unfitness. We all know how much more at peace we are with the rest of creation when we do not rise in the morning depressed by a headache or a torpid liver. Only recently have we learned what intimate relations exist between the presence of adenoids, decayed teeth, and eye defects and the backwardness or obstreperousness of our children. Can one assume that adults do not have similar physical handicaps?

But, at any rate, I have not finished. Scientific Humanism must take other factors into account. Man's career is determined by the sum total of his heredity and his environment. So is the career of every other organism. But man differs from all other animals in his response to mental training. And here is where greater opportunities for improvement exist than in any other department. Heretofore education has been of too uniform a type. We have tried to shape the heads of our youths in a single mould: How much better it would be to analyze their capacities and to train them according to their nature. A similar system should be applied to adults. They have had little aid in determining their fitness for particular vocations; hence, they often do not fit their jobs at all and are unhappy in them. Proper attention to psychological requirements could rectify this situation. These are practical matters. As such it is easy to recognize their importance. Nevertheless, if psychologists worked from now till the day of doom on such special problems, and if their findings were adopted by all the nations of the earth, the essential ingredient of the remedy for the satisfying life would have been missed. Improper conditioning of the mind, to use the behavioristic term, is the cause, I am convinced, of more unhappiness than all other factors combined. This is a

[25]

discovery due to psychological research, yet the situation cannot be amended by laboratory technique. It needs a fundamental change in our entire attitude. The fact is that the members of the human race have lied to each other, continuously and systematically, since speech began. They started on this invidious course through ignorance, perhaps; they continue it, in part, through ignorance; but in the main they persist in it because they are afraid to face reality. Can such a mental condition, rooted in the dawn of human experience, be changed? I am an optimist on the subject. It is changing—changing quite rapidly, as any one historically inclined should know. And those who have made the shift are satisfied that life is richer and more attractive than ever before. The known is not always pleasing, but it is less terrifying than the unknown.

Chapter II

THE PROSPECTS OF THE SOCIAL SCIENCES

by Frank H. Hankins

ONE of the most striking features of the evolution of culture is the sluggish conservatism of thought. This is true even of our very modern culture, which is doubtless the most rapidly changing the world has ever known. As Robert Briffault[1] says: "We are living in a phase of evolution which is known as the twentieth century and stands for a certain achieved growth of the human mind. But the enormous majority of the human race do not belong to that phase at all . . . Twentieth century civilization is cluttered up with living fossils surviving from every barbaric phase of the past, and masquerading as twentieth century people because no attempt has yet been made to ensure that human beings shall wear modern minds as well as modern clothes, and every care has, on the contrary, been taken to provide them with superannuated misfits." Man changes his traditional conceptions of himself and the world with extreme reluctance. This is not entirely due to his stupidity, though the candid student of his tedious climb from animality to modernity finds convincing proof that man is far from being guided by a clear-sighted reason. When one observes that it took our human forebears some hundreds of thousands of years, even by conservative calculation, to emerge from the rough-stone age, and still additional thousands to discover the first use of metals, one

[1] Briffault, Robert, "Rational Evolution," The Macmillan Company, New York, 1930.

[27]

feels warranted in saying that he has possessed barely enough reasoning power to enable him to lift himself out of the animal stage.

Alongside his slender rationality reside powerful mental prepossessions, partly in the form of authoritarian doctrines regarding himself and his world and partly in the form of emotional complexes implanted in youth and enforced by dreadful taboos, back of which are massed the like powerful emotions of the collectivity. In the early stages of his long pilgrimage man had only the tiniest scraps of real knowledge to aid him in his struggle with nature. He peopled the world and even himself with spiritual entities which he endowed with wonder-working power. He made them the causes of events and the controllers of human destiny. That primitive outlook has yielded to the progress of rational understanding, but only after bitter opposition. While several thousand years ago common-sense realism enabled man greatly to elevate his status above what it had been through untold ages previously, the fact remains that the basic assumptions of his philosophy of himself and the world four hundred years ago differed little in essence from those of the primitive view. In fact, in all that touches his own nature and destiny the primitive view still prevails very widely even in the most enlightened countries. If the savage thought his fortune in life was controlled by numerous spiritistic entities who could be influenced by witchcraft and the medicine-man's sorcery, the modern places the same control in the hands of one such entity influenced by prayer, sacrifice, and priestly intervention. Wherein lies the difference between the South Sea Islander's view that infant souls are brought on the waves by mysterious spirits, and the view uttered not long since by a well-known archbishop to the effect that "babies come trooping down from heaven"? The archbishop, to be sure, may have been speaking poetically; one can hardly think that he takes such a statement literally; but his faithful and devout parishioners are expected to do so, and many of them do.

Now the progress of modern science has consisted essentially in replacing mystical spiritistic causes of phenomena by real, natural ones. While there were many adumbrations of the modern realistic view, it was not until the findings of Copernicus, Galileo, and Newton showed the universe in the large to be self-sufficient and self-regulating that a really mortal blow was dealt to any considerable part of the ancient view. A few daring minds thereafter dreamed of the extension of the idea of natural causation to all phenomena. Nevertheless, not much impression was made upon popular thought until Darwin gave a naturalistic explanation of the origin of man and placed him definitely within the system of animate nature.

Since then the spread of scientific presumptions has been rapid. Psychology not only has failed to find any trace of a human soul in the traditional sense; but, just as biology and biochemistry have shown life to be not a thing but a form of functioning or behavior, so psychology finds mind to be not a something but the functioning or behavior of highly organized and coordinated neurological systems. Likewise the ethnologist and sociologist find in tribe and nation purely naturalistic groupings of men, and in moral codes, man-made rules of behavior enforced by group authority and differing with social history and circumstance. Nevertheless, the logic of scientific concepts, of universal causation, and evolutionary naturalism has only begun to penetrate the basic assumptions of popular thought regarding moral conduct and social destiny. While it would be thought wholly incongruous to pray for a change in the location of the heavenly bodies, multitudes still think that prayer and candle-burning will alter the hereditary endowment of a prospective child, cause rain to fall on parched fields, or bring the aid of an omnipotent deity in a murderous war. It is becoming more and more evident, however, even to the man on the street, that progress toward greater personal happiness and a more perfect social organization is intimately connected with the advancement of scientific realism. Authority in religion, and in the morality resting thereon, gives way before a skepticism which queries whether there may not be a way of life that is freer and more abundant, a way illumined by the light of physiological, psychological, and sociological knowledge.

Darwinism was very largely responsible for the development not only of a natural history view of man but also of a naturalistic view of society and its institutions. Obviously, if man is a part of nature, society is also. If society is a culminating product of organic evolution, then the social institutions must be viewed as means whereby societies have sought to maintain and perpetuate themselves in adaptation to their environments. The result was that social scientists of the generation after Darwin busily applied themselves to the significance of the concepts of fecundity, variation, struggle for existence, selection, heredity, and adaptation for the interpretation of social phenomena. It was inevitable that the use of biological analogies would be pushed too far. It was also to be expected that the progress of psychology and ethnology would change the points of emphasis in sociology from the biological to concepts imbued with psychological and culturistic implications. Nevertheless, the biological orientation served to give firmness to sociological interpretation, partly through supplying useful rubrics and partly by removing sociological theory from the deadly grip of a sterile metaphysics.

Among the most widely exploited of these early concepts was that of the social organism. While Herbert Spencer is generally said to have held that society is an organism, his own statement of the matter warrants no such opinion. It remained for certain French and German writers to make such a downright assertion and to attempt to demonstrate it by ingenious reasoning. Spencer, on the contrary, said only that society is *like* an organism because it grows, differentiates, and has systems of agencies which may conveniently be called sustaining, circulatory, and regulatory. He then added that these likenesses are only analogies and that the truths pictured by them were equally true without analogical support. Moreover, being the outstanding protagonist of political individualism, he was very careful to point out that society has no central censorium and hence cannot feel and think as a unitary body. Since the individual is the feeling and acting unit, it follows, he pointed out, that the welfare of the individual is the sole end and justification of social policy and social organization. He expressly stated that society is not classifiable with any known type of organism and that those products of social activity represented broadly by the terms tradition and institutions are genuinely super-organic and constitute a new social environment in which the individual lives, moves, and has his being.

Spencer's view seems essentially sound. When he concluded that, after all is said, the basic and enduring analogy between society and an organism is the mutual interdependence of parts in both, he meant that society may rightly be viewed as a separate and special object of scientific investigation and that it must be conceived as an integrated, complex whole in which are included many delicately interrelated parts. Society is thus something more than the mechanical juxtaposition of a number of individuals. As Spencer indicated, its evolution is attended by an increasing differentiation of parts held together in an even more complex coordination and cooperation of units

[31]

by a perfected integration. These two processes of integration and differentiation, moreover, must keep pace with each other, unless social evolution is to end in a hard and fast solidarity destructive of individual initiative on the one hand, or in an inept anarchy destructive of social power on the other.

While, therefore, Spencer suggested the organismic conception of society, he was fundamentally opposed to a primary conclusion that was soon drawn from the analogy. It was not long before certain socialistic writers seized upon this view in order to strengthen the plea for the reorganization of political and economic institutions so as to give larger application to principles of social solidarity. It is in consequence of a certain inherent opposition of the individual and the collectivity that all the social sciences have struggled with the question whether, and to what extent, the community or the individual constitutes the end or the means. As noted above, Spencer was unequivocal in his answer and, as usual, philosophical certainty was some stages removed from reality. There has been much talk of individual rights and a tendency to set up inviolable principles regarding them, but one observes that these rights are repeatedly invaded by the socially constituted authorities nominally set up to protect the individual in his exercise of them. The basic rule of practice seems to be that the individual has the right to act in his own interest so long as his action does not conflict with what, at the time, the social group conceives to be its own larger interest. Individual rights derive from social approval and depend, for their realization, on the state of group opinion.

Now the reason for this situation would seem to be that the struggle for existence on the human plane has always been carried out on two levels. There is within the social group the struggle of individuals for existence, power, and prestige; and there is the overshadowing struggle between groups. No group ever intends to give its members rights which will weaken it in its struggle with other groups. Moreover, when the latter struggle breaks out in the form of war, or even of economic competition, the traditional and constitutional rights of the individual vanish in a wave of group sentiment which makes group strength and group preservation the primary aim of individual effort. Man is thus, in last analysis, a gregarious animal who, by nature and training, is keyed to an age-old struggle between groups, in which leaders and followers make war upon other leaders and followers, with death or the prizes of life in the balance. At such times the individual is no longer the end but the means; individual happiness and welfare are cast aside, often with a genuine enthusiasm, in order that one's race and blood and all that symbolizes their distinctiveness may survive. At such times society, as an organized, integrated super-entity, becomes a truly Hobbesian Leviathan, absorbing within itself the rights and the wills of its members.

If, then, society is not a true organism, on the one hand, or an anarchistic aggregation of human beings on the other, it must, nevertheless, be viewed as a going concern having an existence of its own quite apart from that of the individuals who, at a given moment, compose it. Tribe and nation endure, though their individual members perish. Not only is this so, but while they endure, they envelop the life of their members in a scheme of rights and duties, a code of legal and moral rules, a set of social values and objects of emotional loyalties that give order, limit, and definite pattern to individual life and personality. It is, therefore, a perennial problem, with which all the social sciences wrestle, to determine when, to what extent, and for what purposes the individual may be freed from group domination to pursue his own wayward and centrifugal tendencies. The political scientist debates the reconciliation

[33]

of liberty with authority; the economist, the merits of individualism versus socialism and communism; the ethical philosopher, the limits of egocentrism and of ethnocentrism, of pacificism and of patriotism; the psychologist, the impulses of self-assertion versus the impulses of self-abnegation and gregariousness. Similarly, the sociologist contrasts the phenomena of individual ascendancy with those of group ascendancy, and dreams of that perfect society in which that social progress, which is so obviously rooted in individual variability and men of genius, shall be as rapid as possible without destroying the delicately balanced and finely integrated relationships among *socii* which give to society the verisimilitude of an organic unity.

Nothing illustrates this integrated character of a society better than the disaster that attends any disruption, disintegration, or unbalance of social solidarity. In our own society we are familiar with the demoralizing consequences of industrial depression for the welfare of millions of citizens. Crime, poverty, physical and moral degradation stalk in the wake of such imperfections of social adjustment. In Russia, a decade ago, the collapse of the established order, in consequence of war and revolution, led to the deaths of millions through disease and starvation. Transportation and communication were upset, unitary control disappeared, and there was a reversion to primitivism and localism unparalleled in modern times. In Spencerian terms, the sustaining, circulatory, and regulatory systems broke up, with the result that the elaborate cooperation necessary to the life of the people as a whole collapsed, destroying the delicate web of relationships that bound city to city and city to village and forcing local units back to a basis of self-dependence. While such extreme cataclysms are convincing evidence that orderly continuous variation in social evolution must be—in terms of life values—enormously superior

[34]

to explosive mutations, and while they reveal how small is the actual control of society over its own development, the processes of orderly change are themselves so little understood as to produce the endless debate between individualism and collectivism. The material progress of the last century and a half seems convincing evidence that a considerable degree of individualism is justified by its fruits. However, the student of society cannot overlook the fact that such rapid change leaves in its wake an appalling amount of poverty, maladjustment, and degeneracy. The greater the speed of change, the greater the strain upon individuals to make adjustment to the ever new conditions. The system of free enterprise and voluntary inter-relationships has resulted in an almost dizzying rate of invention and economic improvement for society as a whole; but it also produces alternating periods of prosperity and depression with unemployment, crime, suicide, and nervous collapse for many. These evils may be looked upon, in large part, as the price we pay for liberty and its dynamic effects. Like all selective processes in nature, those of a competitive society are rough and ready, and do damage along with benefit. The problem of the social engineer is to preserve the latter while reducing the former.

Spencer's organismic analogies thus made it clear that society is a special order in nature whose structure and evolution must be studied from a "natural history" viewpoint. Darwin showed that man, and hence society, are products of evolutionary processes. They do not, therefore, escape the action of that struggle, selection, survival, and adaptation characteristic of all living things. On the human plane all these take on a somewhat special character. This would be expected in view of man's special qualities. As related to this particular matter, these special qualities would seem to be his superior brain power and his natural defenselessness.

[35]

This latter attribute should be as obvious as the former. Man has neither tusks nor claws, neither thick hide nor furry coat. He is devoid of specialized anatomical structures for offense or defense. These his ancestor did not need so long as he was a tree-dwelling, fruit-eating animal. But the transformation of this ancestor into a land ape was a momentous step, for otherwise he must have remained immured in the tropical forests and could never have begun his conquest of the globe. But with this descent from the trees he must soon have perished had he not acquired distinctive ways of maintaining himself. Moreover, the human infant, as John Fiske long ago pointed out, is not only extremely helpless, but reaches self-dependence at a later age than any other creature in the animal kingdom. Human survival, therefore, required mutual aid or cooperative effort, first, between parents in the rearing of offspring; and second, between fellows, in the chase and in warfare. This situation has given rise to an enormous literature discussing the questions whether man has a parental instinct, maternal and paternal, and a gregarious instinct. Suffice it here to say that the parental patterns, however much they may differ from group to group in consequence of cultural differences, seem everywhere to be based on clearly marked instinctive tendencies. Even male man everywhere reveals a tender regard for children and an easily developed willingness to labor for their maintenance. In the earliest stages of human culture, except in extremely favorable habitats, it seems difficult to account for the preservation of mother and infant in the absence of an instinctive tendency of the male to remain close at hand to ward off enemies and to assist in maintenance. It is true that he might have done so out of a desire for companionship or as a consequence of habit and the memory of moments of sex gratification. But whatever theory one may adopt, the facts remain that the family unit is found on the very lowest levels of existing culture where there is every evidence of its primeval character, and that prehistoric man developed habits of maintenance and preservation that enabled him to spread to habitats of every kind. Finally, we may note here, that so important did the family appear as the primary unit of social life and racial reproduction that man very early surrounded it with the most inviolate mandates of his highly variable moral codes.

But man's natural defenselessness has affected his social evolution in another profound manner. It was an essential condition of his gregarious habit. There is obviously some question whether man has a truly gregarious instinct. Moreover, the easy explanation of human behavior as due to this or that instinct is distinctly less popular to-day than it was even a decade or two ago. Still further, it is not quite certain that man began the group habit of living in consequence of an instinctive tendency rather than as a result of an intellectual perception of its utility. It seems probable that in many habitats the individual family, or even the solitary person, could have managed to exist without the advantages of group cooperation (especially after the acquisition of crude weapons, if we may suppose such to have been possible in the absence of associated life); but that would have been a precarious and ill-adapted mode of life. Not only from the standpoint of food getting, but also from that of reproduction, life in a horde would have been superior to seclusion. I think we are warranted in holding at least that man acquired an instinctive predisposition toward the acquisition of gregarious habits.

Such gregariousness as man developed, however, was not that of the bovine herd. Religion has long and often urged man to submission and humility, but has seemed to succeed only with the weaker members of the species. In order to

[37]

understand human history as an expression of human nature, one must see man as an extremely complex and highly variable creature basically much more akin to the Canidae than to the Bovidae. It is no mere unique and accidental phenomenon that the dog was the first domesticated animal and that there is a more sympathetic and intimate understanding between man and dog than between man and any other animal. The thesis advanced and admirably sustained by Professor Carveth Read[1] that man is descended from a wolf-ape which acquired a taste for meat and developed the habit of hunting in packs is at least plausible. Such a theory, in any case, is in harmony with those conditions of life essential to the development of longer legs, agility, erect stature, intelligence, language, and social life. Moreover, it is in harmony, not only with the universality of the hunting-pack pattern in social life, but also with characteristic psychological traits expressing themselves therein. Conspicuous among these traits are those of leadership and following, an order of precedence, rivalry, loyalty, lust of the chase and the thrill of killing, cunning and strategy. The leader must prove his superiority and must be followed with unswerving loyalty. The individual must cooperate with the pack and hence must be subservient to the interests and moods of his group. Nothing is more common at all stages of human evolution than the formation by strong and aggressive men of bands of followers for purposes of hunting, fighting, robbing, plundering, intimidating, and otherwise controlling and regulating the lives of weaker or less well-organized persons. Among our Nordic ancestors, to live by robbery and plunder was honorable and meritorious, and the *comitatus*, or organization of chief, lieutenants, and followers, was the unit out of which have been forged

[1] READ, CARVETH, "The Origin of Man," Cambridge University Press (The Macmillan Company, New York), 1925.

feudatories innumerable. Those patterns of social organization, made so familiar and so conspicuous by the well-developed feudal order, are exemplified in the Australian tribe, in the Indian hunting or war party, and in many phases of modern life, such as governmental institutions and political parties, business corporations and racket gangs, ecclesiastical institutions and university faculties.

As stated above, man is more wolfish than bovine in temperament. The instinctive tendencies of the herd animal are manifested by huddling together, running in the same direction, heightened suggestibility in times of danger, uneasiness in isolation, and obvious pleasure on being restored to the herd. Man manifests these traits to a considerable degree. He loses himself in the crowd which he readily follows and whose mental attitudes he thoughtlessly shares; his high suggestibility is repeatedly manifested in his collective gullibility, especially in times of group danger; he finds solitary confinement a dreadful punishment and rejoices in the restoration of free association with his fellows. On the other hand, the herd animal is non-aggressive, prone to flee from danger, and manifests aggressiveness only under severe provocation; it does not find in the chase, warfare, and killing either a continuous interest or deep emotional gratification; it knows nothing of strategy, courage, or indomitable will in attack; it feeds quietly, slowly, and peacefully and not greedily, enviously, and pugnaciously. Man shares with the dog and wolf those traits in which they differ from the bovine herd animals. These traits are by no means an adequate representation of his highly complex nature, but it is easier to understand him and his history in the light of such a biological interpretation than in that of the softer view assiduously cultivated by authoritarian myth and given credence by human vanity. He is, in fact, the fighting animal *par excellence*, but he fights with greater gusto in a

[39]

group than when alone. Like other creatures he will fight when alone to preserve his life and, in some cases, his honor and dignity; but in groups his courage and confidence are enormously elevated, his sense of power and adequacy become inflated, and he becomes aggressive, domineering, and blood-thirsty. In view of his historical record, the devoutly cultivated myth that man was created in the image of a loving God and endowed with a sweet, kindly nature only a little lower than that of the angels, must be looked upon as a most extraordinary compensatory rationalization. Every page of human history is spotted with senseless and unnecessary bloodshed, crimes of violence and ferocious brutality. Warfare and crimes of robbery and violence, individual and collective, seem to be ineradicable from human relations, just as are also the existence of Tammany Halls and Ohio gangs in politics, and exploitation and monopoly in business.

Is it not most truthful, and hence wisest, to recognize at the basis of these facts deep-seated elements of human nature? And what is more reasonable, from the standpoint of evolutionary biology, than to find these elements rooted in the age-long struggles of our omnivorous ancestors combining as best they could to win a short and brutish existence from a raw and none too friendly nature? Karl Pearson,[1] from a comparative study of the femora of man and other primates, concludes that the ancestor of man was not like the shrinking tree shrew that runs to the tree tops on the approach of danger but "heavy in build and violent in character," "agile in motion, slender in his proportions, gracile in his bones, and dexterous in his flight from possible foes," "a violent fighter and a ravenous feeder." "We owe more than half our trouble to-day to this ancestry. Is the brutality and violence of man to-day a fall from

[1] PEARSON, KARL, "Side Lights on the Evolution of Man," *Eugenics Laboratory Lecture Series*, No. 13, p. 6, 1920.

a higher estate, or an atavism, a relic of a violent past?" Such a view, in so far as true, should not be looked upon as degrading to man. It does not alter his true nature, but is rather an effort to see him as he is. Human nature is no doubt deeply affected in its individual development by its social environment; moreover, civilization appears to grow less and less favorable to war and violence. But in all our longing for a pacific world, we must not overlook the advantages as well as the disadvantages that flow from man's courageous fighting qualities. In their absence it is inconceivable that he would have risen to the mastery of nature, conquered sea, desert, and mountain, or even performed brilliant feats of daring and endurance unparalleled elsewhere in the animal kingdom. Moreover, if through some mutational change or the slow effects of natural selection, he should everywhere become meek and docile, the course of social evolution and the character of social organization would be profoundly altered. They would be different; but whether they would be more satisfactory is a question.

Returning, then, for a moment to the gregarious tendency, we may note that the predominance of the group struggle for existence over all else depended upon and reinforced powerful sentiments of group solidarity. There must have been a time when groups were so small and territory so vast that conflicts between them were relatively infrequent. Such a condition still exists along the arctic fringe of human habitation. Elsewhere the competition for the best hunting grounds led very early to traditions of war, diplomacy, government, and peace. Man met in himself his most redoubtable enemy so that conflicts between groups became the supreme tests of racial quality, social organization, and cultural equipment. Man's native equipment made the cultivation of the art of war easy and inevitable. Accustomed to hunting in packs, the relation

[41]

of leaders to followers, of authority to obedience, was a part of his original equipment; his courage, aggressiveness, vanity, and thrill in combat were ready to hand for military exploits; and his intelligence enabled him to utilize with unexampled skill those stratagems used by the wolf or dog pack in running their prey to earth, such as attacking *en masse*, or by surprise and stealth, surrounding, intercepting, ambushing, and similar movements.

With warfare a developed and immemorial habit, social organization and cultural evolution took forms adjusted thereto. Authority or government within the tribe or nation, the gradations of rank and occupation, the status of male and female, standards of honor, honorableness, and truth, and even religious institutions and practices are very largely adjusted to the exigencies of warfare. Tribal loyalty became the basic virtue. The distinction between what the late Professor Sumner aptly called the "in-group," or ourselves, and the "out-group," or others, became sharp and incisive. Internal unity has always been enforced by the most powerful mandates of government, morality, and religion, in order that group strength might not be dissipated. The closer and stronger the enemy, the sterner has this internal discipline been. In consequence of this "struggle of races," as it was called by the eminent Austrian sociologist, Ludwig Gumplowicz, there arose that ethical dualism which has been the bane of all philosophical and universalized ethical systems. In relation to one's own gang, whether tribe, political party, or business group, one must be loyal, honest, truthful and steadfast, charitable and helpful. In relation to the "out-group" one becomes meritorious in proportion as he is deceitful, treacherous, lying, vacillating, cruel, and destructive. Moreover, whatever in the form of habit, custom, agency, or institution exemplifies one's own group becomes, in one's own esteem, vastly superior to the crude, ludicrous, or debased

[42]

ways and possessions of the "out-group." One's own government, morals, and gods become sacred and inviolate, while those of the "out-group" are wicked and contemptible. Civilized man, to a very limited extent, outgrows these age-old attitudes, though warfare revives them even in those academic circles where scientific candor and objectivity are cultivated as the highest virtues. Moreover, they are sometimes explained as mere expressions of human vanity; whatever in any way typifies one's own group is to some extent identified with oneself. Consequently, one's own egoistic satisfaction requires that all such things be highly esteemed. This explanation is doubtless true so far as it goes.

At the same time, we may see in this spirit of clannishness both a consequence of, and an aid in, the group struggle for existence. "The strength of the wolf is the pack; and the strength of the pack is the wolf." What heightens individual courage and confidence increases the group chances of success. Confidence in the strength and superiority of one's race, and tradition and faith in the omnipotence of one's gods, are assertions of the group will to live and to conquer. In their absence, courage weakens and degeneration presages defeat and elimination. The study of the decay of tribal societies, both in this country and in the South Seas, reveals the profoundly significant fact that the crushing of tribal confidence by white supremacy destroys in the individual even the will to live and to reproduce.[1]

Finally, one further word may be hazarded regarding this "struggle of races" which has for so many thousands of years dominated the human scene. It has inspired men to their most heroic efforts, stimulated invention, forced repeated modification, and even the annihilation, of culture patterns. It has always been an essential element of empire

[1] Rivers, W. H. R., "Essays on the Depopulation of Melanesia," Cambridge University Press (The Macmillan Company, New York), 1922.

[43]

building, and the great peoples of history have all been inveterately warlike during the periods of their greatest efflorescence. If, however, war was a factor in their creation, it was also the source of their destruction. In other words, ability to survive in the supreme arbitrament of war has been the final test of fitness of peoples and institutions. Nevertheless, war seems rightly to be rated as the greatest evil of the modern world, the chief obstacle to the construction of a world order based on ideals of peace and universal material prosperity. Much as such a goal may be devoutly hoped for, it is still doubtful whether the human animal is as yet sufficiently domesticated and ethically perfected to make it attainable. Moreover, it is also doubtful whether, if it were attained, we have sufficient knowledge or sufficient means of social control to enable us to preserve and perfect racial soundness and social institutions in the absence of the crude, expensive, wasteful, and often apparently senseless destruction provided by the final verdict of war.

We noted above that two distinctive traits of man were his superior brain and his natural defenselessness. It is obvious that man's greatest distinctiveness lies in his brain power, for this is the chief source of that super-organic or psycho-social environment which he alone has produced. Superior brain and natural defenselessness would seem, in the evolutionary process, to be set over against each other. As the former develops, the latter becomes increasingly feasible. Man's special superiority in the former has enabled him to develop artificial means of protection and adaptation. While man has undergone some modifications of skin texture and color, hair texture and color, eye color, nostrils, and other anatomical and physiological racial differences in consequence of mutation, isolation, and environmental selection, adjustment to habitat by cultural devices is enormously more facile and efficient than by the biological processes. Man has sought to escape death and soften the rigors of an omnipresent natural selection by improvement in the arts of life. Culture is a buffer between man and nature. It enables him to ward off his natural enemies and improve the quality of life, while also increasing his numbers. In consequence of it, and the above-mentioned collective aggressiveness, he is the only mammal that has spread to all parts of the globe by its own initiative.

Culture may be most briefly defined as inclusive of all habits, attitudes, skills, beliefs, arts, laws, customs, morals, and material possessions acquired by an individual as a member of a society. It is a product of three universal human traits; namely, (1) the tendency to live in groups; (2) the powers of associative memory (capacity to learn or to profit by experience); and (3) the capacity for articulate speech. With these, man was equipped to develop implements to supplement the hand in the chase and war; clothing for ornament and warmth; houses for protection from the elements; language and communication; moral, political, and juridical codes to govern the internal relations of associates; myth, magic, religion, and science to govern the action of the forces of nature. This culture thus constitutes man's answer to the problem of existence. It is his unique and unrivaled mode of adaptation to nature. Everywhere it stands as a buffer between a society and destructive forces, whether these arise from the physiographic environment, from hostile human groups, from the breeders of dissension within the group, or from that transcendental and mystical world of spirits which man imagined all about him.

This view has the merit of bringing every aspect of culture within the range of the natural. None of it is miraculous or mystical in source and essence. Even those parts of it especially designed to explain or to deal with what are conceived as mystical entities and experiences come within

[45]

the range of the natural conceived in psycho-social terms. Culture is a measure of man's adaptation to nature. It must thus be looked upon as in no sense divinely created or inspired, but as an expression of human needs and human capabilities seeking a solution of the multiple problems of adjustment to physical and social environments.

Numerous attempts have been made to list these needs and capabilities. Not long since, it was the habit in the social sciences to make long lists of instincts and thus to explain all that man does or has done as the expression of an instinct. Nowadays this sort of explanation is a little like explaining the origin of life as due to "spontaneous generation." This does not mean that man may now be supposed to be entirely free from native tendencies to action, but rather that the existence of an instinct must not be assumed but must be demonstrated. Man forms habits so easily that, although these may be built upon inherent neural patterns, he may, for all social science purposes, be looked upon as a creature of habits instead of instincts. This view is obviously more in harmony with the facts because human behavior is so highly variable from culture to culture that the theory of inherent instinctive patterns of behavior is of little sociological use. There are, however, certain "drives," urges, or primary needs which may be looked upon as universal sources of human activity. Among these may be listed food, sex, bodily security, activity itself, and egoistic satisfaction. One might include here gregarious and parental urges. Some of these drives are primarily physical, or neuro-muscular, others are primarily neural, or psychic. They are implicit in the cultures of men at all stages of social evolution, and hence are satisfied in a great variety of ways. It was from man's efforts to satisfy these needs that culture first began. Impelled by them and guided by his feeble intelligence, which deluded him more often than it led him aright, we see man stumbling and zigzagging down the corridors of time, only occasionally, and at long intervals, emerging from the darkness of superstition and reliance on unseen powers into the light of intelligence and reliance on rational cause and effect.

Social origins are lost behind the veil of the past, but the general scheme of culture is the same everywhere. There is first some degree of practical knowledge and development of the industrial arts. Here are included kinds of food, clothing and shelter, tools and techniques, means of transport, rights of property and barter, and forms of personal service. These have primarily to do with the satisfaction of physical needs, especially food and security. There is, secondly, some plan for controlling sex expression and securing the perpetuation of the species. Here are included the sex mores, marriage customs, the family, and the rights and duties of kinship groups in relation to each other. There is, thirdly, language and communication. There is, fourthly, an enormous range of ideas and practices regarding the nature of the world and man's relation thereto. Here are included myth, magic, religion, medical lore and, latterly, scientific knowledge and research. A fifth aspect of culture includes all those manners and customs regulating the private relations of individuals, such as ceremonial forms, private morals, voluntary clubs and associations, play and sports. Sixthly, we may include under ideas and practices regulating the public relations of individuals those ethical rules that are believed essential to group welfare, and all juridical and political customs, forms, agencies, and institutions. There is, seventhly, all expression of aesthetic emotions and interests, such as personal adornment, drawing, painting, sculpture, music, and architecture. And, finally, the arts and traditions of war and diplomacy. These constitute the universal culture complexes. Each and all vary from culture to culture, though all are found in every one. Moreover, they are

[47]

inclusive of all that men anywhere have devised by way of cultural device.

Though origins are irretrievably lost, we are not without some insight into some of the principles and conditions that have guided social evolution. Early cultures are purely and simply adaptations to environment. From very early stages this environment included not only physical habitat but also the spirit world. Later the political environment became highly important, while with advanced cultures we see adaptation to the conditions of world trade and world politics. We may conceive primitive man, like our-selves, trying, by trial and error, various ways of satisfying his wants, always seeking, with the least effort, to avoid pain and increase pleasure. The more successful ways tended to be adopted and so to become the folkways of the group. One could, without distortion, describe this as a process of selection among variates. That a full set of culture traits, as outlined above, was attained in the very remote past is shown by the fact that the most primitive races known have cultures which, in one or more respects, are highly elaborate. Since culture is man's device, he cherishes it; among savages, even apparently trivial matters are often imbued with a ceremonial rightness which makes departure from them dangerous and sinful. The primary reason for this is that a large part of the folkways become mores— that is, they become imbued with a quality of sacredness because believed essential to group welfare. Age gives authority among primitive minds; and the more ancient a custom is, the more it is believed entitled to sacredness. Back of the mores are placed awe-inspiring and fearsome mandates of the gods with which primitive man peoples his world. Henceforth they become inviolate; their right-ness and adequacy are no longer proper subjects for critical thought. Every culture thus resists, in greater or less degree, invasion by alien cultural elements, partly because it cultivates in the group a strong sentiment of superiority and partly because it tends to become a homogeneous whole. Alien elements, therefore, set up cultural conflicts which tend to affect all parts of the cultural scheme. Isolation deepens cultural homogeneity, fixity, and distinctiveness. For this reason, certain primitive peoples may well have lived for thousands of years with little or no cultural change. The mores set the patterns of approved thought, of "right thinking," at all times and places; few, even among modern well-educated persons, ever think or feel except in those terms inculcated in them by their mores.

If these reflections are sound, one seems warranted in drawing two inferences. One is that age is not a useful criterion of the validity of a doctrine or belief. The more ancient such tenets are, the closer they approach the original innocence and ignorance of the race. The other is that the doctrine of relativity applies to all that relates to human institutions. Ethical rules relating to sex, family, property, public agencies, and private rights and duties are all relative to time and place and must be judged, not by absolute principles, or by the light and darkness of our own culture, but by their fitness as means of adaptation in the culture where they are found.

Slow as cultural change has been during most of human history, there nevertheless has been change. It results from whatever produces an internal crisis or external contact or conflict of one group or its customs with another. Those eight or ten periods which have produced so-called civilizations since the first arose in the ancient east have all been periods of contacts multiplied through trade, immigration, and travel, and conflicts in the form of war and conquest. Such conditions set up processes of selection, elimination, and recasting in the whole cultural scheme. Probably the basic changes are those in industry

[49]

and war. These touch closely the perennial needs for food and security. Economic activities are less imbued with inviolate taboos than are most others; they are more nearly commonplace and hence more readily modifiable in the light of experience and rational thought. Methods of warfare are subjected to the selective action of competitive value. Changes in economic life especially tend to exert a modifying influence on all other aspects of the associated culture. There is a constant "strain toward consistency"[1] among the mores. New ways are brought under old sanctions in the same way that Christianity devoutly approves both a holy war and world peace, or first opposed the charging of interest and later embraced the capitalist economics, only to be transformed thereby.[2]

These transformations of culture often have a quality of rationality attributed to them that does not properly belong to them. The judgments of a social group no doubt represent all the powers of reason it can muster, but the premises of thought are mainly derived from the mores, and the action of reason is guided by the overpowering emotions associated with ethical, religious, and philosophical principles. The facts which would expose reality are obscure, while only proximate ends can be envisaged. The resulting adjustments are, therefore, hit and miss affairs. Abundant illustrations may be found for modern times in Herbert Spencer's writings—especially in his "The Sins of Legislators"—showing that the unexpected results of laws often greatly exceed the expected; in Bertrand Russell's exposition of "the harm that good men do"; and in the thousand and one instances in the evolution of western culture where the distant consequences of an

[1] SUMNER, WM. G., "Folkways," Ginn & Company, Boston, 1915.

[2] TAWNEY, R. H., "Religion and the Rise of Capitalism," Harcourt, Brace & Company, New York, 1926; MAX WEBER, "The Protestant Ethic and the Spirit of Capitalism," trans. by Talcott Parsons, Charles Scribner's Sons, New York, 1930.

invention or other cultural change were wholly undreamed of by the participants. Would a Susan B. Anthony and Elizabeth Cady Stanton not turn in their graves if they could be made to realize that they helped to mother birth control and trial marriages? A conspicuous current case is the Eighteenth Amendment. Cultural evolution is thus a blind stumbling. Great civilizations have repeatedly arisen only to disappear, while the peoples that created them knew neither why they rose nor why they fell.

Nor can we now speak of such matters with any confidence. However, the realization that realistic cause and effect are universal in scope gives some ground for optimism as to the future. We have in organized scientific research a culture trait that is unique. The Greeks came very close to its discovery; had they once realized its significance, they would have changed the course of civilization during the last two thousand years. We, on the other hand, have at length placed science among the mores and made it the final judge of truth and validity. The social sciences, however, make slow progress toward that positivity which is essential for clear-sighted control. This is mainly due to two facts: first, the difficulty of extricating social research from the biases and taboos entrenched in the economic, political, and ethical mores themselves. Scientific objectivity is immensely more difficult to attain. Second, social phenomena are immensely more complex than those investigated in any other field. They result from an almost infinitely variable interaction of physiographic, biological, psychological, and cultural factors which make the detection of general laws extremely difficult. Moreover, the applications of science to social life necessarily, at present, have reference only to immediate problems and scarcely at all to the remote future. For example, medicine, sanitation, and psychiatry tend to give an increasingly scientific quality to a vast range of humanitarian activities.

The ultimate consequences for racial quality, and hence for social evolution, are only dimly perceived.

A final difficulty in the achievement of that social control which is the goal and final test of scientific validity is that public opinion determines both the introduction and the effective working of new truths. It is not easy to achieve in popular thought that degree of rational insight necessary for the rapid readjustment of political and ethical institutions. The growth of rationality in these fields is necessarily slow and takes place by a tedious transformation of popular views through suggestion and imitation of one social class by another. The gradual emergence of science to a position of chief authority among the mores is probably the most significant cultural change taking place at present. The warfare between science and theology is real and epoch-making. They represent two directly opposed modes of explaining phenomena, the animistic and the realistic, the creationist and the evolutionary, the irrational or supra-rational, and the rational. The future of our civilization depends on which one succeeds to the place of unquestioned and universal primacy among the criteria of thought and emotion. From this point of view the popularization of science and the increasing attachment of popular thought to scientific concepts and attitudes are bright omens of a new and better era.[1] In the light of present knowledge, we seem warranted in saying that the future of our culture is dependent on learning *how to accomplish by scientific methods* that elimination of the unfit and ill adapted which now is accomplished by natural selection on both the biological and the sociological planes. Nature's methods are haphazard, wasteful,

[1] For an interesting statement of what has already been achieved in the direction of a new texture of social life, see page 1402, "The Science of Life," by H. G. Wells, Julian Huxley, and G. P. Wells, Doubleday, Doran & Company, Inc., Garden City, N. Y. 1931.

and cruel, however beneficient and effective in the end. Artificial control in the light of tested knowledge is direct, economical, and humane. Nevertheless, the thoughtful student of the human epic does not anticipate that man will be able fully to control his social destiny. He aspires only greatly to decrease suffering and to increase happiness. When Hobbes declared that the life of man in a state of nature was "solitary, poor, nasty, brutish and short," he epitomized a remote age in the past. We may likewise imagine, under the salutary ministrations of science, a remote age in the future when human life will be sociable, comfortable, wholesome, humane, and enduring, and when the aesthetic values will be all pervasive.

Chapter III

THE RENAISSANCE OF PSYCHOLOGY

by JOSEPH JASTROW

MODERN psychology derives from the application of the
scientific method to the study of human nature;
it studies objectively the mental mechanisms and motives
and their organization; but such "objectivity" includes
the capacity to reflect upon experience. To the twentieth
century mind this statement seems as simple and convincing
as that nature includes mental nature. It is, however, the
issue of a long, irregular, and difficult evolution, a slowly
accumulated heritage of the ages. We may distinguish
three trails[1] of interest in man's concern for knowledge
about himself; they appear in ancient as clearly as in recent
stages. The earliest is a combined religious, cosmic,
philosophical expression—a *psychosophy*, or more simply,
soul-lore. By that inclusion, psychology, while a new
science, has an ancient history. Soul-lore dominates in
early as in primitive reflection; it mingles dramatic invention with inquiry, making myth the antecedent of science.
In this approach the study of the mind is incorporated
with philosophy. There arises a body of doctrine philosophical in temper but with increasing contributions of
psychologically minded philosophers; it characterizes
the course of inquiry from Aristotle to Descartes, and from
Descartes until the advent of the scientific concept of
psychology—a half century ago.

[1] As suggested by M. Dessoir, "History of Psychology," The Macmillan Company, N. Y., 1912.

[54]

The psychology embodied in it had to be sifted out of the philosophical context by later historians. Stratton has been able to compile a "Physiological Psychology among the Greeks." Gardner Murphy has set contemporary psychology in a perspective of the preceding centuries. Psychology, as we recognize it, had to await a general renaissance of the scientific method and the objective attitude, and to await particularly a more searching biological insight. Besides psychosophy and psychology, there is the practical interest in the affairs of mind, for which the term *psychognosis* has been suggested. The diagnosis of human character, the judgment of men and motives, ever with a hope of control or prediction of fate, has made the "mental" man the proper study of mankind. It is quickened by the experience with minds disordered seeking cure, no less than with strange experiences demanding explanation, hinting at a hidden order of "psychic" world.

This threefold body of interests is persistent: (*a*) the place of mind in the world at large; (*b*) the mechanisms of the mental activities and their association with bodily functions centrally; (*c*) the varieties of character, the vicissitudes of the human scene, the direction of mental affairs, including the relief of ills that mind is heir to. Gradually the scientific claimant came to be recognized as the legitimate sovereign, absorbing the philosophical and the practical dominions in an inclusive rule.

The historical sequence appears also in terms of concepts of *phusis* or nature, by virtue of which things come to be and are maintained. How nature is regarded determines all else.

The story of science centers about the nature of nature. It encounters a strong preconception as to what nature should be to conform to human desires or beliefs; it meets with the tendency to imagine forces beyond nature, or miracles setting aside nature; it emerges with an objective,

[55]

patient study of the environment and the organism. It comes to recognize in nature forces beyond human influence or control, except by the route of comprehension; it comes to recognize man in all respects as part of nature, with affiliations deep and significant with other organisms, no less than with endowments unique but not unrelated to minds of other kinds. Nor can man by any inspiration step out of the scheme of nature, or overcome his limitations, save by that same instrument of understanding that makes the human estate possible.

The history of science thus considered, and psychology centrally, presents a pre-naturalistic or an anti-naturalistic phase and stage; it clings tenaciously to a super-naturalistic intrusion; slowly it emerges to a consistent naturalism. It was by such irregular course that the psychologist became a naturalist. The renaissance of psychology was dependent upon the advent of naturalism. The doctrine of the soul, with its ethical and religious implications and cosmic place, is both pre-natural and anti-natural; the magical-mystical development in primitive and advanced religions goes over to the super-natural, and reappears in so modern a claim as a "meta-psychics." Yet they all fuse from the outset and mingle their products variously and ever with a core of true psychology proceeding upon "natural" observations of the phenomena of the mental world. Naturalizing psychology among the commonwealths of science has proved to be a tortuous process, consummated only in our own generation.

The modern student is interested in understanding how psychology in the scientific temper views and studies the life of the mind; that is the central purpose of this contribution. It can include but a limited survey of the data of the science, selecting by emphasis discoveries that determine principles of interpretation. As already indicated, that issue—the scientific viewpoint—was long delayed through the occupation of the field by concepts and interpretations that to us seem extraneous, irrelevant, crudely speculative, unsupported, and unsound. Their logic seems as alien as their findings. As we go back but a few intellectual generations, we seem to be in a foreign land, where our way of thought can find few points of contact and no sense of illumination, no resting place.

The picture of the mind as we know it to-day is scarcely recognizable; so much is out of focus, so much distorted, so much intruded, so many vital features lacking. An historical retrospect seems like an excursion into a land of errors and phantoms. This may in a measure be true of the antecedents of all science, but applies in larger measure to psychology, because the mind was so open a domain, approachable from so many avenues—a circumstance which accounts equally for the wide reach of its modern applications. All these accumulations of tradition and obstacles to understanding had to be removed or corrected by truer approximations before the heritage that we accept as naturally as steam and electricity and radio came to be ours. The layman of to-day is as familiar with psychology as with physics or chemistry. Man is living in a psychological age; he has developed an intimate interest in interpreting human behavior, as well as the physical resources of the modern world.

It is worth a moment's delay among the antecedents of psychology to gain a truer sense of the recency and significance of its establishment. If we go back to Descartes (1596–1650)—that great mind of a great period, who devised analytical geometry, hailed Harvey's discovery of the circulation, and had a high place among the philosophers—we find bodily behavior explained by "animal spirits," animals reacting as machines, but human animals behaving by a totally different "rational" nature, in that

[57]

man has a soul. For that entity he had to find a local
habitation and a name. He was advanced enough to look
for it in the brain, which, unfortunately for his notion of
unity, was divided into two hemispheres and symmetrical
throughout; so he selected the pineal gland, which is
single and near the center of the brain, for the place of
honor as the seat of the soul. Descartes, as psychologist,
likewise considered the "passions" or emotions, conceiving
them in similar mechanistic fashion, analyzing them as
intellectual calculations in terms of pleasure forces, or as
actions and reactions of attraction and repulsion. What
the gland actually was, he could not know. What is
pertinent is that he did not think it necessary to inquire.
He actually "proved" his own existence by the fact that
he could think about it: *cogito, ergo sum*. He thought he could
get at the bottom truth about things, and particularly how
we can have knowledge of them, by emptying his mind of
all presuppositions, then examining them critically, and
demonstrating their validity. It has been neatly said that
he took all his mental furniture out of its house, examined
it, and put it back again. To us this seems learnedly
futile; yet it represents the level and procedure of the
"psychology" of three centuries ago.

Leibnitz (1646–1716), the leading philosopher of the
following generation, regarded the relations of mind and
body as regulated by a "pre-established harmony," with
neither as cause or effect of the other; the two clocks,
body and mind, were set once for all to keep the same time.
In grading mental processes he came upon the fact that
some were obscure and unconscious, coming from within
the body, others clear perceptions directed outward—a
discerning observation but not followed to its rich
possibilities.

Hobbes (1588–1679) was mainly a pleasure-and-pain,
profit-and-loss psychologist entering the field from the

economic and political approach because he needed such data to portray social behavior. He recognized the major instincts and their influence on behavior, yet with no further insight into their relations than that the primitive baseness of human nature required management in the interests of social worth.

Locke's (1632–1704) position is of peculiar importance because he brought into prominence the central problem of the senses. Unless we understand where the mind gets its material, the finished product will be unintelligible. He was prompted by just the same empirical interest that later developed the psychological laboratory, in which the functions of eye and ear and the tactile senses formed the first objects of inquiry. There was nothing in the mind that had not been at one time in the senses. Mind worked with an equipment; the products of thought required a knowledge of the sensory instruments. He had equally the educational interest in training the senses to make better minds. Thus he distinguished between the primary sense data and the secondary qualities that developed upon them. We know time and space and all the physical relations first in their sensory phases. These we build up into perceptions which make intelligence possible.

Through all this runs the notion of a simple lowly source of behavior (reflexes) and the recognition that the child, beginning with a blank slate (*tabula rasa*), slowly grows to the adult mental stature, the route proceeding by way of the perceptions in sensory terms to ideas in abstract form. This empirical correction was the more necessary to keep thinking realistic because of the strong tendency to describe it all in more abstruse, unanalyzed terms. Characteristic of this approach—a favorite one in Teutonic thought—is to proceed at once to posit a set of "faculties" such as memory, reason, will, which in turn constitute the "soul," and to describe the varieties of

[59]

experience and endowment in these naïve terms, not yet realizing that memory must have an organic counterpart, and that reason is but a name for a reaction toward perceived relations. Of like import was the "associationism" that flourished in Great Britain, which explained, far too abstractly, how the world of ideas was composed and directed—forgetting how closely it was modeled on the world of things as organized in words.

"Faculty" psychology had a strange sequel when Gall (1758–1828), a distinguished anatomist, was led to conclude that he could judge from the shape of the head what "faculties" the owner possessed in marked degree. By this completely fallacious method he discovered an impressive array of "faculties," from philoprogenitiveness to tune and number, and located each on the cranial surface. The unscientific phrenology which he founded had a flourishing career, but disappeared when the true nature of cortical functions was established.

All this is a mere sample of scattered observations—an eclectic array of the piecemeal, mosaic fragments of psychological knowledge filtering through a variety of philosophical and scientific pursuits, unassembled, unrelated, and unorganized. The relationing of it all into one scheme of mental life, with an origin and a growth and a mutual illumination—all so obvious to our insight—did not enter upon the horizon of these forerunners of psychology. For this the general intellectual advancement was as yet insufficient, the interpretation of nature lacking in profitable clues.

The establishment of the theory of organic evolution in the middle of the nineteenth century completely reconstructed the biological sciences; it gave a new key to the varieties of life, placed structure and function in a new light, and projected a genetic version of organic processes, as applicable to the highest as to the lowest organisms.

This biological renaissance determined the advance of psychology. Psychology, as we know it, was impossible before Darwin, not because of the evolution concept alone, but because of the confluence of several rapidly growing streams of knowledge and of method at this period.

The renaissance of psychology was launched with the impetus of the greater wave of evolution. Leaving that for the moment, we must focus upon the experimental approach; the two have in common the objective attitude; and if we include the naturalistic viewpoint, the prolific fertility of modern psychology is traced to its headwaters. Negatively, experimental psychology is a protest against armchair assumptions and subjective elaborations in the systematic manner of philosophy or the eclectic temper of literature. Positively, it is a reform in logical method, deriving remotely from the Baconian emphasis on induction—though this alone was not sufficient to direct psychology into scientific channels. The realistic pursuit, forsaking speculation and clinging close to observation, found a radical expression in the positivist program sponsored by Comte, with its direct influence upon English pragmatic thought.

That movement has its center in the problem, persistent from the Greek to the Darwinian naturalists, as to how the affairs of body and mind are related and integrated. When Wundt, the first mind to be inspired by the vision of a comprehensive scientific psychology, came to formulate it in a massive volume, the parent of five-foot and fifty-foot shelves of modern psychological texts, he properly called his systematic treatise "Physiological Psychology." He is known as the father of experimental psychology and the founder of the first psychological laboratory (1879). Though the chair he held, true to the older tradition, was titled "philosophy," he followed the clue of the times in focusing upon the physiological approach.

[61]

The first chair in Physiology was established as late as 1833.

Until mental functions, from low to high without exception, are considered in their biological origins and counterparts, a scientific psychology lacks central unity. That is a fundamental relation confirmed by evolution; for bodily and mental functions emerge and evolve by the same set of forces. The mind's heredity and the body's heredity are one. We are not thus curtly disposing of the mystery of mind, not asserting that thoughts and emotions may be reduced to patterns of behavior in the nervous system. But since in many ways we know more of mind than of body, the demonstration of the reality of the connection between the two remains a guiding principle throughout, though we can apply it but partially.

The connection can be suggested in a paragraph; it can be covered only in a volume such as C. Judson Herrick's "The Thinking Machine." The advancement of physiological psychology from the days of Wundt to those of Herrick is equally striking. A machine the mind is, but of its own type of organization, with only secondary analogies to man-made mechanisms. There is no more hesitation in speaking of one integration of "brain" areas as the organ of consciousness than in speaking of another as correlated with digestion, and another with sex; and both these "physiological" functions may induce elaborate and disturbing "conscious" representations.

The hierarchy of the central nervous system conditions fundamentally and throughout the hierarchy of the mental activities. The simplest member of the series was termed a reflex:[1] Here lies the basic unit pattern of response. And

---

[1] It is as true of scientific as of personal names that their baptismal status gives no indication of future career. "Reflex" arose from an outdated notion of actual reflection—as of a ray of light from a mirror—but was physiologically conceived with the appearance of the neurone theory and of neuronic integration. Sir Michael

with its mention we may as well face the aggressively contentious issue of behaviorism which holds that at the beginning was the conditioned reflex, and there has been nothing since. In every logical sense the copyrighted "behaviorism" for which John B. Watson claims proprietary rights, filed (presumably in the Library of Congress) in 1912, has no place in a serious and responsible discussion of the origins and tenets of modern psychology. It is included with apologies; but its omission might be misunderstood. The claim that scientific psychology began when the term "consciousness" was dropped and "behavior" substituted, is as false as it is trivial and captious. The turn away from the emphasis on the "rational" processes, detachedly considered, toward the inclusion of all varieties of mental behavior, is a gradually increasing by-product of the naturalistic trend above outlined. It appears in successive waves, the first at least a hundred years ago. With few if any exceptions all modern psychologists are behaviorists.

To charter the word "behaviorism" for one's personal vehicle is an impertinence in one sense or the other. No man of science has any time or taste for the querulous game of logomachy. As for the radical and far-reaching conclusions which Dr. Watson regards as "behavioristic," they belong to the story of the mental vagaries of able minds. It is because his views for a time found a limited following that their mention is warranted as a perturbing incident in the orderly progress of modern psychology.

The study of the reflexes, including the limited order of "conditioned" reflexes with which the serious work of Dr. Watson was concerned, emphasizes the basic importance of primitive, lowly behavior patterns. But the com-

---

Foster's "History of Physiology, etc." (Cambridge University Press, Cambridge, 1901), takes the story to the point at which the recent reflexology of Pavlov continues it.

plexity of even these in their actual integration is the impressive contribution. Of all the "primitive" psychologies of recent years, the glandular psychology is the most illuminating. The glands of internal secretion are notable for their varied influence on growth and behavior; the externally secreting glands are no less important, with their distinctive connection with the several divisions of the autonomic nervous system. We have a rich documentation that mind and body are of one organic "conditioning," which is more correctly a "dispositioning" appearing in temperament, and through temperament in career. That gland deficiency may precipitate gross mental incapacity, that excessive glandular functioning induces emotional and "rational" disturbances—all this gives "physiological psychology" an intimacy and scope of meaning available to no previous generation. Such suggestions as that the new (biologically) pons or great cable of connection between big and little brain, makes possible the amazing skill of the human organism in acrobatic posture and manual dexterity; that the overgrowth of the neo-pallium, displacing and replacing the rhinencephalon or "olfactory brain" of lowlier creatures, conditions the human rational stage of behavior—these are but glimpses of the endless script of the affairs of mind, legible in the partially deciphered characters, gross and fine, of the nervous system.

Such knowledge affords data for a version more profound than is yet available of the neurological romance. It is this plot that found popularity in "Why We Behave Like Human Beings." In recognition of this directive approach, our present era may be said to find its focal expression in the neurological concept of behavior. Such a conjunction is vital to modern psychology.

In continuing the genealogy according to the records, we find the first experimental studies in the field of sensation modeled on physical concepts. E. H. Weber discovered the variable sensibility of the skin by showing that a pair of compass points pressed on the finger tips, separated by an eighth of an inch, could be felt as two, but had to be separated three times as far on the back of the hand to get the same effect. Thus it is found that the fine discrimination of the human hand, along with the delicate guiding movements of trained muscles, is the physiological basis of skill. We carry a complete psychic map of our sensory and motor equipment. That is the biological interest in these early discoveries. Their academic pursuit led to the formulation of the psycho-physic law, the real interest of which is that it proves a psychological relativity. By endlessly patient lifting of weights to determine which was heavier as the differences were reduced, it was determined that an additional ounce in a pound felt as much heavier as an additional two ounces in two pounds. Such investigation played a leading part in Wundt's program and led to studies of the reaction times of mental processes, to studies in attention and association, in perception and apperception. The psychophysical interlude absorbed the academic mind; but its net issues were not momentous. It neglected physiological interpretations though it enriched the experimental technique, which was destined to become an important factor in the equipment for psychological research. It was essential that psychology should become experimentally minded. How the mind gets its data has since become the indispensable starting point for all psychology.

The genetic viewpoint has a closer relation to the interpretations that go to make the psychology of to-day; but the progress of thought that is its history has diverse origins. Its two foci are the child mind and the animal mind; its body of doctrine constitutes comparative psychology. Studies in animal behavior are as old as zoological interest; they were not systematized until close to the experimental period. They were preceded by anecdotal accounts of rare intelligence in animals, at times fantastically interpreted with a humanized slant; but analysis was lacking. Animal psychology is entitled to a place of esteem in the renaissance of psychology on several counts. It offers the most strictly objective procedure. It studies an organism as a whole; it surveys varied types of mind, all specialized, but on far simpler patterns than the human; the conditions of study can be fairly well controlled, and the observations repeated and confirmed. It illuminates the process of learning, for animals are trainable; it indicates more clearly what is inherent and what is acquired; it makes definite the concept of stimulus and response.

It is not accidental that the behavioristic (in the liberal sense) and the *Gestalt* psychologists should find their fundamental data in animal behavior. The results are as important in showing what animals cannot do as in demonstrating their intelligent capacities. That the human response should be interpreted as far as possible on the animal pattern has become an established principle; but it is equally important to set forth the superiorities and differences of the human solutions. We know more of how we came to behave like human beings from the study of the contrasts and limitations of animal behavior than from any other single source. The psychological interest grows as we approach the nearest of kin and study the "Almost Human" behavior of the anthropoid apes, which Yerkes describes. Anthropoid psychology, in the main, is a contribution of one or two decades. It proceeds upon the combined clues of physiology, evolution, the growth processes, and the differential functions of the highest nerve centers. It forms a brilliant example of scientific psychology and brings psychological concepts into the biological field. Dunlap, aiming at a rigidly scientific conception, regards psychology as psycho-biology.

A comment upon the place of "animal psychology" in the psychological renaissance of the twentieth century is particularly in point. There the biological trend in interpreting the domain of mind appears most commandingly, and from there it has spread to other problems and has determined, above all else, the general point of view from which the entire field of mental phenomena shall be approached. Recent advances in the concept of the units and varieties of behavior shape the working principles of the modern psychologist; they apply from reflexes to personality. Without the illumination of studies in animal behavior, the correct approach to the study of human behavior would have been impossible. By no expansion of the movement that founded the Wundtian type of laboratory would the illuminating views of human response have been possible. The convergence of the experimental method and the biological viewpoint made modern psychology.[1]

The older tendency was to humanize animal behavior, to apply crudely, and with prejudice, types of conduct that arise from awareness of motive and process and goal to the limited and differently oriented capacities of animals. The fallacy of that procedure has become obvious. The reverse approach constitutes the distinctive merit of the behavioristic position. Thorndike (about 1890) formulated experimental methods appropriate to the animal problem; Lloyd Morgan had previously exposed the weakness of the humanistic approach; Loeb introduced the mechanistic concept, perhaps too rigidly; Jennings showed the vari-

[1] A concise history of the stages by which animal lore, from ancient times to our own, became a branch of scientific psychology, is presented in a convenient form by C. J. Warden, "A Short Outline of Comparative Psychology," W. W. Norton & Company, New York, 1927.

ability of responses even in the lowest organisms; Watson developed the technique of record and analysis; Holmes gave the mental behavior its proper place in the zoological setting; Yerkes gave the subject the most comprehensive interpretation; the *Gestaltists*, notably Köhler, demonstrated the overlapping stages of human and animal performance. None the less, as Yerkes emphasizes, the naturalistic interpretation that studies animals in their native habitat and records their native solutions of problems, free from human guidance, furnishes a major clue. Experiment may be too refined and too artificial; a broad behaviorism includes both.

The ancient problems thus appear in transformed aspect. The distinction between instinct and reason loses its rigidity; in all cases we deal with behavior patterns. Animal learning proceeds by trial and error; the unlearned factors in response are instinctive. Instinctive trends are modified by experience, both spontaneously and in response to human training. Behavior, with insight into the relation between means and ends, coming not like "trial and error" with slow improvement and correction but with the sudden abandonment of an old method and an adoption (after reflection) of a new one, appears clearly in the more intelligent specimens of the anthropoid apes— perhaps not elsewhere. The interplay of heredity and environment appears more clearly. There is in most animal, as well as human, behavior "conditioning" in the more liberal sense and equally "dispositioning." Their variations make way for the problem of individual differences which, in the human setting, becomes paramount.

Child psychology has a different set of antecedents. It goes back to Pestalozzi and Froebel and the "kindergarten" movement—all with a strong interest in directing the earliest educational procedures. Infant life provides the closest approximation to objective conditions and supplies a genetic picture at the earliest stages. Darwin himself was among the early pioneers in writing the "Biography of an Infant"; but it was not until the elaborate technique of Gesell that the possibilities of this field were realized. The study of the growth processes is of paramount importance in all the applied arts of guidance. Genetic psychology is still largely in the making; for it includes decline as well as growth; it makes plain the importance of the age factor in all responses. Its many implications appear in other connections.

The profound influence of child study—root and stem, branch and blossom—derived from the psychological study of childhood and the unfolding process, can hardly be included, or yet omitted, from this survey. It has made the movement called "progressive education," which recognizes the freedom and spontaneity of the stages of child growth, which places emphasis upon emotional development, which makes central the learning to live with other people, by way of character growth and normal social reactions. The specifically mental development takes its proper place. Discipline has acquired a newer and a richer meaning. The problems of adolescence are recognized; abnormal developments are avoided. Parental guidance assumes new aspects. Parallel with this insight is the specific study of the educational procedures; for obviously the equipment for the practical life in the actual environment of the human scene is indispensable in the formative period. The analysis of the speech processes has made great strides. Such an impediment as stuttering is realized as a personality defect even more than as a lack of coordination. Educational psychology takes its place among the important applications of psychological principles. The social aspect of the child in the family life has been freed from the too drastic intrusions of tradition and moralized methods negligent of the complex relations

[69]

of character growth. The place of sex enlightenment and the relations of the sexes at all stages is frankly faced. The nursery school has appeared as a direct expression of the foundation of control upon natural development. Learning by doing, creative exercises at all stages, self-government, protection of the child's spontaneity, the avoidance of too much regimentation, the stimulation of individual aptitudes, all enter into the program.

It becomes apparent that modern psychology has assumed its present status by a composite set of advancements, partly reflecting those of the general story of science, with its enthronement of logical procedure and refinement of experimental and observational technique, but particularly dependent upon the data and concepts of the collateral life sciences—biology, physiology, neurology. It has been equally enriched by the clinical sciences of medicine, specifically of psychiatry. Mental abnormalities are of profound significance for normal psychology; they do not stand apart from but are framed in a single picture of the mental life. The extreme departures, whether remediable or due to original inadequacy, have naturally claimed the attention of the psychiatrist. But in this reference psychology has given even more than it has received; for within the twentieth century—with sporadic roots in earlier studies—psychiatry has become psychological. No estrangement seems to our modern insight so illogical as that of the mind in health and the mind in disorder; no rapprochement has been more helpful to both parties in the scientific league of mental studies. The data of abnormal psychology have always been available, even insistent; the concepts necessary to their understanding were lacking.

The early interpretation of insanity was demon possession, which the medicine man or priest might exorcise—though in the face of idiocy he was helpless.

States of ecstasy and trance were sought for prophesy and religious rites. Strange notions originated as of the influence of the moon, perpetuated in the term lunacy. The mentally disordered seemed dehumanized. The humane conception begins with Pinel (1791), who first dared to release the shackles of the maniac and to consider the true sources of disturbed behavior. In the absence of a clue to the vicissitudes of mind, such belated theories could for a time command attention as those of Mesmer, claiming an "animal magnetism" paralleling the physical force, but streaming through his person and by that route inducing and allaying what we plainly recognize as hysterical manifestations. Such a concept as hysteria (retaining in its etymology the false notion of a connection with the uterus) in its modern renaissance compasses a large range of aberrant phenomena. Abnormal psychology has been naturalized into a confederation of disciplines conferring insight into human nature.

How much of this enlightenment is traceable to the Freudian invasion is not easy to determine. What may be termed the clinical phase of psychology was certain to arrive, even if there had been no Freud; it was written in the course of the emergent psychology, whose genetic stages we have been roughly and eclectically following. But the fact remains that the essential Freudian concept had a revitalizing influence similar to that of the Darwinian contribution. Yet this, too, has its antecedents in sporadic points of insight from Galen on: that the sufferings of mind were not only closely related to bodily infirmities and disqualifications, but also reacted upon them—a tenet now formulated as the psychogenic principle. It took on a novel Freudian form when hysterical lameness or blindness—a bodily symptom—was referred to a mental mechanism—a subconscious repression, expressive of a motive to avoid, to shut out, to escape a situation too harrowing

[71]

to be borne. It was the discovery, or at least the novel formulation, of a new order of mental mechanism.

The concepts of the Freudian system can hardly be reduced to a synopsis. Yet, as a guide to the account that follows, it may be stated that the purpose is to explain the motivation of conduct in all its varieties. The instinctive urges, the sex urge most of all, supply the energy and give set and direction. The urge to power, mastery, regard, is the ego urge. When these meet with opposition, there results conflict, frequently below the conscious level. Such unconscious or suppressed tendencies may come to expression symbolically, in dreams; they may lead to a neurosis or maladjustment. Character traits arise from sublimations and compensations for such frustration. The systems of emotionally tinged ideas become complexes when elaborated, and become fixations when expressed in a personal relation, notably that of child to parent. Childhood influences dominate in the total personality, and the family relation sets its course. Much of it is in sex terms. Psychoanalysis is a method of revealing buried complexes of ideas from which maladjusted individuals suffer. Phobias and obsessions thus take their root and grow. The same processes are at work in the primitive unconscious and are responsible for totem and taboo, for myth and fairy tale, all of which carry a Freudian plot. Social repression in terms of a censorship plays a large part. Upon this outline has arisen an elaborate system of interpretation calling for hypothetical divisions of the ego as the nucleus of the strivings which make the personality and account for its desires and behavior.

The Freudian psychology, a piecemeal structure with slight architectural unity to begin with but growing by irregular additions from clinical experience, offers an inviting opportunity to illustrate how variously a specific advance extends the psychological insight. Beginning with

[72]

the interpretation of an hysterical symptom (such as an inability to move a limb, yet with no true paralysis; eyes that do not see, yet with no real blindness; ears that hear not, yet without impairment), it was extended to a variety of fundamental neuroses. It was then made a clue to the interpretation of dreams, a subject almost outlawed from recognized psychology and relegated to the pseudo-psychology that continues the folklore products of cruder ways of thought. Still later it was applied to everyday lapses, mishandlings, forgettings, in all of which the same factor of motive gave a new meaning to incidents and reactions seemingly accidental. And finally it was enlarged to a comprehensive formulation of human motives, furnishing an insight into character traits, emotional disturbances, social relations generally. It introduced a new determinism. Still more importantly it projected a psychology from above, complementing that "primitive" reflex psychology, which is a psychology from below. Both are phases of an inclusive neurological concept of the sources of behavior. The question of its validity in principle is not the same as that of the correctness of its details. Its contribution is permanent. We shall never return to a pre-Freudian any more than to a pre-Darwinian outlook.

The manner in which Freud, and especially his followers, have elaborated the details of the drama of life, has aroused violent protest within and without the medical profession. It is denounced as unscientific and speculative, as basing large conclusions on slender (and subjective) evidence, using such treacherous devices as symbolism, reading remote meaning into the obvious, ignoring well-established diagnoses. Its reception by psychologists is likewise divided; many are Freudians, but with large reservations. Freudianism has proceeded far more as a cult than as a science, and if it persists in its present temper, its status will continue to be uncertain.

[73]

The Freudian doctrines include principles, mechanisms, and applications. Each is entitled to appraisal on its own merits, though the structure stands as a whole. In the first division, the exploration of the subconscious, the distinction between logical thinking in terms of reality and fantasy as pleasurable wish-thinking, the keystone position of the affective life, the importance of childhood experiences, the far-reaching determinations of the sexual life, fit in with the progressive trends of modern psychology, as well as with the close parallelism of normal and abnormal manifestations. The mechanisms of repression, substitutional outlets, compensations, rationalizations, fixations —all leading to the concept of the complex with its emotionalized force—have been fairly well incorporated into a motivation psychology.[1] In the applications, psychoanalysis overshadows all else. Its value depends intimately upon the perspective of principles and the wisdom and tact of the practitioner. The Freudian clue to character analysis, the possibility of a psychobiography, the emphasis upon the proper psychic weaning of children to and beyond adolescence, the Freudian factor in the psychoneuroses, however treated, apart from the interpretations of dreams, trance states, and lapses, indicate the possibilities of the Freundian instrumental aids.

But approached closely, the weaknesses of the construction are conspicuous; and one can sympathize with Dunlap's classification of the entire system as unscientific and mystic, while declining to agree with his dismissal of it on that ground. Had Freud been content to indicate the generic plot of the human drama, he would have remained on safer ground. He proceeded to outline its details. Sex played the leading rôle and dominated the minor parts as well. There developed the "family romance"

[1] See L. T. TROLAND, "Fundamentals of Human Motivation," D. Van Nostrand Company Inc., New York, 1928.

in which parent-child relation leads to tangled complexes; genital details, sexual threats, traumas, perversions, invade the normal course. As a consequence the psychoanalytic procedure makes later neuroses the issues of such incidents and seeks their unmasking. Both as principle and practice, this phase may be rejected as a speculative and a cultist development, unfortunately appealing to a popular vogue, and catered to by disciples lacking a sober logical training. There remains a core of doctrine that, once separated from its extravagant deductions, may be assimilated with progressive psychology. As for the treatment of mental disorder, a sound psychiatry is not disturbed by a school of practitioners claiming exclusive jurisdiction in the name of an innovator in one branch.

The correction of the Freudian doctrine has proceeded by way of rival dissensions. Jung is clearly the master mind in the psychoanalytic movement, and is sensitive to its philosophical and cultural implications. He introduced an experimental technique in the detection of complexes by delayed, peculiar, and distorted associations; he set forth that the will to power and self-expression is a far more pervasive and dominant motive than sex alone. He is devoted to symbolism even more than Freud, but defends it as a literary embodiment of psychic elaboration. The effect on practice is not distinctive, though it leads to a more liberal diagnosis. Alfred Adler finds a single clue to life adjustment in the avoidance of an inferiority feeling. A dominant mechanism is compensation. He advocates an acceptable form of psychoanalysis, since adjustment proceeds in terms of the total personality. Jung calls his system "analytical," Adler styles his method "individual" psychology. The intelligibility of the Adlerian procedure brings it nearer to the common-sense approach, but adds little to its scientific status. It readily falls to the level of an eclectic, somewhat evangelical guidance.

The permanent value of Freudianism in the renaissance of psychology is not clear. It has contributed a new vocabulary to general usage, has influenced literature and morals and conversation, has reinterpreted social prejudices and personal failings. Its scientific status can hardly be appraised until the cultist phases of the movement have disappeared.[1] Freud himself, in his later years, has turned to still more speculative formulations of the hidden depths of the personality, speaking of the super-ego and the "Id" and even of a meta-psychology embracing the system of strivings of which the complexities of life consist. But whether a structure or a superstructure, this interpretation of human motivation and personality must make terms with the psycho-biological foundations. Dr. Rivers is almost the only one who has attempted to determine the evolutionary significance of that stupendous growth from primitive instincts to the elaborately rationalized emotional conflicts which characterizes the modern mind, often to its undoing.

It is because we live in an increasingly complicated world and have become mind-conscious in desirable and undesirable ways, that the Freudian type of psychology has made such a wide appeal. The problems of emotional and intellectual adjustment have become acute; we seek happiness by tortuous routes. The big brain has unbalanced its subordinate yet more fundamental partners in an integrated enterprise.

With as many ill from mental disturbances as from all bodily ailments combined, the safeguarding of the mental economy by intimate knowledge of the higher motivation

[1] Fortunately there has appeared, at this critical juncture, a most discerning critique of the movement: "The Structure and Meaning of Psychoanalysis," by William Healey, Augusta F. Bronner, and Anna Mae Bowers, Alfred A. Knopf, Inc., New York, 1930. For the first time there is available a systematic account, with an able interpretative comment, of the significance of the entire movement following upon the epochal innovations of Freud.

[76]

routes will remain an essential part of education and social direction. That conviction has found expression in the world-wide movement for mental hygiene. Suddenly the human scene has become a clinic, and sick humanity—the mentally lame, halt, and blind—groping their way uncertainly in a world too complex for their limited powers of adjustment, cry out for relief. To answer this appeal exceeds the combined resources of the psychologist and psychiatrist, aided by law, church, and social institutions. We seem to be living in a psychoneurotic as well as a psychological age.

There is one phase of psychopathology having such fundamental influence upon the integrity of the social structure that it alone would demand the development of a "psycho-clinic"; namely, that of crime. Crime, we are prepared to admit, is a social-economic manifestation. Economic pressure, playing the game of self-interest too roughly, using ingenuity to avoid labor, short-circuiting the currents of profit, exploiting the feeble, is a sufficient temptation to human nature to make eternal vigilance and effective education the price of public safety. None the less the prevalence of crime is as intimately a question of maladjustment, neural inadequacy, and psychopathic trends. All that society demands is crime insurance. The nature of the criminal instincts is a matter for psychological diagnosis; treatment must follow psychological guidance. As a remedial measure the present system has failed. Crime shades into delinquency, and delinquency into minor maladjustments. Crime is so largely a juvenile phenomenon, a question of management of unruly urges, that it stands as a constant menace to the security of the maturing process. Criminal psychology[1] covers every aspect of society's dealings with offenders. As a clinical problem,

[1] See Gross, H., "Criminal Psychology," Little, Brown & Company, Boston, 1911.

it is that of the management of low-grade individuals with weak resistances and strong temptations. The influence of the approaches of modern psychology appears at every phase of the social provisions for the guidance and treatment of criminals. The modification of criminal procedure presumably will proceed slowly as a consequence of the heavy cost and inefficiency of established methods. The institution of juvenile courts, which provide for intimate case histories and examinations of the offenders under the guidance of a psychiatrist who sits with and advises the judge, the establishment of institutions like the Judge Baker Foundation in Boston for the study of delinquency, form practical evidence of the growing confidence in the psychological approach to the problems of crime.[1]

By such route we reach an extension of the psychological disciplines in a significant direction, that of the social structures which men have instituted to express and satisfy their mental needs. The data of social psychology are ancient, their interpretation modern. Sociology grew to an independent estate by staking a claim at the frontiers of several sciences where their boundaries failed to meet. Psychology thus includes not only mental processes but mental products.

With his characteristic insight Freud recognized early that if the mechanisms whose issues he was then tracing in the neurotic field had any claim to universal significance, that claim must be justified by discovering the issues of similar motivations in the cultural products of mankind. In the same spirit of a psychoanalytic explorer, he found complex expressions in wit and humor, in art and literature, that could be translated into the Freudian formulae. Wit was letting the cat out of the bag, or a self-betrayal; art

[1] See HEALEY, W., "Mental Conflicts and Misconduct," Little, Brown & Company, Boston, 1917.

and literature disclosed Narcissistic trends. Narcissus, like Oedipus, grew out of the unconscious artistic following of clues deeply imbedded in the psychic nature; Greek and Roman fantasies achieved a renaissance in psycho-neuroses.

But even more directly pertinent is the tracing of these trends in the anthropological institutions of totem and taboo, the protections and prohibitions by which the primitive mind had safeguarded its unconscious recognition of psychic values, culturally expressed. Freud's contributions to psycho-anthropology are incidental. The anthropologists, from Tylor to Levy-Brühl and Boas and Radin, and from Spencer to Frazer and Crawley and Malinowski and Briffault, have laid the foundations for a psychology of primitive mentality that has important consequences in the interpretation of human cultures, low and high, simple and evolved. To cap this edifice with the social psychology of our own economic, political, and social management, was the completing stage. Meanwhile the folklore movement had gathered inexhaustible material for tracing the story of mind in its popular expressions. The psychology of the crowd, from Le Bon to Trotter and Conway and Martin (who applies the Freudian concepts), sets forth alike the products of gregariousness and of the collective integration in which communities have their being. Social psychology has become a specialty of large proportions; it, too, dates from the application of evolutionary concepts to the social structure. It would have been impossible without the collateral contributions of the several developments that came in the wake of the renaissance of psychology in a brief half century. It brings evolution once more to the forefront.

Applied psychology may be briefly surveyed. Its *fons et origo* lies in the modern pragmatic attitude, which asserts that if psychology is an actual working clue to the life of the mind, individually and collectively, it must leave the academic confines of a specialty or the narrow loyalties of a "pure" science—an ideal still cherished in early experimental days—and launch out upon the world of human relations and practical enterprises. Psychology must "work" and be put to work.

Distinctive in this development is the story of mental testing—distinctive because it arises from so central a problem as the nature of intelligence and was guided by so practical an aim as to provide a measuring scale applicable to a great variety of uses. It arose from the desirability of detecting the subnormal children, incapable of following the program of the schools fitted to normal requirements. It has come to be an instrument employed in education, in psychiatry, in the case histories of the dependent and delinquent classes, and in industrial occupations. Intelligence is a flexible term; it has its basis in the general level of development, the neuronic adequacy and finer organization of the higher brain centers, and in the development of these highly specialized structures. We are not musical and mechanical and acrobatic and graceful and socially adaptable and apt at analysis and deep reflection, by virtue of quite the same patterns of development, the same supply and organization of engrams. And that basic fact itself gives rise to an engaging division of modern psychology of which Sir Francis Galton is the founder—"The Psychology of Individual Differences."[1]

Human variability, like the prolongation of human infancy to which John Fiske first directed attention, is an asset of inestimable value. Upon it the diversities of employments composing modern civilization have been developed. It is the trend of civilization to make the finer differences count. The guiding principle is that we all grow mentally, and perform more or less successfully, by virtue

[1] Surveyed in a volume under that title by R. S. Ellis, D. Appleton & Company, New York, 1928.

of the special abilities supported by the general level of our intelligence.

The nature of intelligence and of the variety of special aptitudes can but be referred to as conditioning the problem of the job; for it is far from so simple a matter as avoiding square pegs in round holes, or the reverse. Human aptitudes are of as many varieties as the geometric constituents of design. The industrial psychologist has appeared upon the scene, and personnel officers are attached to large corporations. Their service depends upon psychological findings.

That modern industry sets problems for the psychologist is a fact that need not detain us long. It may lead to a lack of proportion between intrinsic interest and the accident of economic worth, as is conspicuously true of the psychology of advertising—a "shop" specialty that happens to engage large financial investments. The true components of that equation are the set of desires and motives on the one side and the appeal that will direct them into the spending budget on the other. Advertising may add needless evidence to the familiar fact that a fool and his money are easily parted; and it may offer minor illustrations of the fluctuations and uncertainties of taste. The suspicion that much of it is economic waste is growing among such psychologically minded economists as Stuart Chase. As a topic it has far less intrinsic interest than the psychology of prestige or of persuasion, of which it forms a minor chapter.

By contrast, the psychology inherent in educational procedure or in industrial efficiency justifies the expenditure of time and technique in its study. Lost motion, whether in laying bricks, assembling a machine, or teaching children the rudiments of arithmetic, geography, history, manual arts, is a matter important enough to demand a proper analysis of its causes and a search for its remedies. The formation of proper manual habits is an indispensable step to right living in which the voice of psychology is claiming and receiving a hearing.

There has been slight attempt in this eclectic survey to apportion mention to importance, or to aim at any completeness of inclusion. Much has gone into the making of modern psychology that is omitted from this sketch. The purpose has been to convey a set of impressions that may compose into a picture—a motion picture—selecting significant moments in the story of how modern psychology came to be, together with excursions into characteristic scenes within its comprehensive domain.

To the same end a selection of general questions that have a material bearing on viewpoints in psychology may be added.

For one of these a place might have been found earlier in the exposition. It is equally appropriate in this connection, in further illustration of a problem of reflection persistent from Greek days. For the complement of the precept: "Man, know thyself," is the injunction: "Man, know thy world." How we come into relation with the stimulations to behavior, what it is that we react to, as well as how we react, are the problems of a psychology that has achieved the dignity of an independent name, not easily anglicized—"*Gestalt* psychology." The *Gestalt* is a concept of correction; its problem is the fundamental one of the terms in which the organism reacts to the environment; what it particularly corrects are the limitations and ignorings of a strict behaviorism. The convenient stenographic formula of the mechanism incorporated in a unit of behavior is stimulus and response, S-R. It is a label, not an explanation. As the strict behaviorist interprets it, it is a formula omitting the essential ingredient. It should read S-O-R—Stimulus + Organism + Response—and the clue to the relations of S and R lies wholly in the O. It is the complete organism that determines what shall act as a stimulus and what form the response shall take. Because the clue to behavior resides in the organism, some psychologists have preferred to speak of this approach as *organismic* psychology. Why the behaviorist should choose to ignore the Hamlet of the play is not wholly clear; it certainly simplifies the plot, but it denatures the drama. It may be through dread of the later consequences, for Hamlet soliloquizes, which is a form of introspection and a manifestation of consciousness—both anathema to one adhering to the act as the sole datum. As Köhler, the leading exponent of *Gestaltism* suggests, the strict behaviorist (such as Weiss and Meyer) is determined to model the S-R reaction on a physical pattern, thus robbing it of its vitality. "But between the two terms of the sensorimotor circuit there is more *terra incognita* than was on the map of Africa sixty years ago"; and it is, as we have seen, to the successful exploration of that psycho-neurological province for sixty years that we owe the present rich knowledge of the psychic organism.

Gestaltism properly emphasizes that we react always to the *total situation;* it is not a stimulus, not a single condition, but a configuration that induces a reaction. The workman reacts to the whistle by going to work in one setting (time) and by quitting in another; and he doesn't react at all if it isn't *his* whistle. The phenomena are so complicated that it is often a problem to determine just what phase, what combination, what *Gestalt* sets off the reaction. The ingenious technique devised to this end is well illustrated in Köhler's "The Mentality of Apes"; for *Gestalt* psychology has contributed importantly to the analysis of animal behavior. It is equally in line with the genetic sequence. For as we grow, we react to the (apparently) same situation more completely, ever to more complex *Gestalts*. This applies to emotional as well as to intellectual situations. *Gestaltism*[1] thus supplies a significant correction of a fundamental problem. It is in accord with progressive psychology and, while making no claim to provide a program for the entire field, introduces a point of view comprehensively applicable.

In historical perspective the Wundtian period of psychology, following the clue of its antecedents—of Herbart particularly—was *structural* in its conception; it became *functional* by absorption of the biological viewpoint. A better description of this change of direction is from static to dynamic psychology,[2] for the latter emphasizes an organic force that runs through all of our drives, whether these drives are due to instinct or to motive. The word "mechanism" is misleading. Dr. Healy even chooses the term "dynamisms" to designate the Freudian factors in directing behavior.

It was indicated at the outset that psychology considers the mechanisms and motives of behavior and their organization, giving them ever a naturalistic interpretation. Certain phases of psychology, therefore, will focus upon the mechanisms or dynamisms, as in the neuro-psychological problems of perception and response; others will center on the higher motivation processes, as in the Freudian scheme; still others will concentrate on the organization. *Gestaltism*, for example, is a psychology of organization. All three phases always combine. Psychology is a unit, not compartmentalized; it merely divides its problems in order to make them accessible, and to make them applicable to the varieties and needs of human nature and of human institutions. The rivalry of the psychologies is more

[1] See KOFFKA, K., "The Growth of the Mind," Harcourt, Brace & Company, New York, 1924; KÖHLER, W., "Gestalt Psychology," Horace Liveright, New York, 1929.

[2] See WOODWORTH, R. S., "Dynamic Psychology," Columbia University Press, New York, 1918.

apparent than real;[1] yet a still greater fusion of complementary methods and programs is essential to further progress.

A most comprehensive problem of psycho-biology is that conventionally summarized as the apportionment of behavior to heredity and to training, to nature or nurture. That mental heredity must follow the clues of physical heredity is certain. With such a novel and definite advance as followed the rediscovery of Mendel's experiments, there arose the hope that the secret of the mental inheritance had been found; and in a sense, it has. From peas to humans, the laws of genetics apply: the genes, with their definite mode of organization and of hereditary transmission, are as responsible for qualities of brains as of the skeletal tissues; they hold from mice to men.[2] Like genes breed like genes; but the drawings are made in a composite lottery. The particular genes which appear in our own make-up, and from whence they came, remain uncertain. In one sense, therefore, the problem has been as much complicated as solved. Life remains a chance—our chance in parents and ancestors, and theirs in us. Just how far genetic procedure may be applied in analyzing the mental dowry is still a vexed, though a clarified, issue. The general trend of it all is clear: mental heredity is real though variable; it is carried in the same mechanisms, genes, chromosomes and all. Yet the world will ever have to depend on chance selection for its favored specimens, whether modernly controlled by the route of eugenics or not. Education has its set limitations.

Any psychology with an hereditarian emphasis is certain to proceed differently, in some phases, from one definitely

[1] My address on this topic before the International Psychological Congress (1929) appears in *The Scientific Monthly*, November, 1929.

[2] See Herrick, C. J., "The Brains of Mice and Men," Chicago University Press, Chicago, 1926; Wheeler, W. M. ,"Foibles of Insects and Men," Alfred A. Knopf, Inc., New York, 1928.

[85]

placing environmental influences as the major determination. The "behavioristic" position (in Watson's narrower sense) has staked everything on environment. Apparently the argument is based on the "conditioning" of the original "conditioned reflex," as exemplified in the Pavlov salivation response in the dog, in which, after salivation is induced both by the presence of food and so arbitrary a stimulus as the sound of a gong, the physiological reaction is set off for a time by the sound alone—an interesting observation, but one with limited application. Another support is found in the small number of built-in reflexes in the human infant, ready to function at birth, and the ease with which a stimulus may be directed to one form of response (fondling) as opposed to another (shrinking). Whether an infant fears or fondles a dog can fairly well be directed if one starts right; and the same is true in terms of fearing snakes and mice. All this is familiar and finds its explanation in the genetic unfoldment of the reaction paths. Yet upon this slight foundation has been reared the monumental structure of a psychology that assimilates all learning to "conditioning," ignores original bent and limitations, and magnifies training to a commanding influence in disregard of specific endowment.

The confusions and ignorings involved in this position are too many to be exposed in a sentence; they illustrate to what glaring contradictions a strict environmentalism would lead. As a fact the trainability and plasticity of the central nervous system—increasingly noticeable as the higher centers or patterns of integration are involved— bring it about that we humans are singularly free from "conditioning," and in that freedom lies the clue to our learning power. Occasionally, perhaps frequently, we may succumb and may even shape the fortunes of habit to our service; but all that is incidental to the "unconditioned" areas of the productive, flexible mental life. What writer

[86]

would not be delighted to have himself so conditioned from early childhood and by later training that whenever a bell rang or a chord in G major were struck, he would respond by productive composition! A glorious but futile dream, and how inconvenient if the signal came inopportunely!

There is no salvation in conditioning. Indeed the more helpful concept for that process comes from the Freudian angle and receives the name of "fixation," an undesirable and emotionally overloaded attachment or complex or idiosyncracy. It there assumes the complexion of something to be avoided by keeping attachments flexible and in perspective. Yet equally is that an environmentalist conclusion. As a fact there is no phase of the mind's activities, above the most rudimentary, which is not subject to the combined influences of nature and nurture, of dispositioning and conditioning, each rather flexibly interpreted. Even the relative constancy of the intelligence quotient is not wholly a matter of endowment, but results from a relative equality of environmental stimulation; hence the statistical regularity and the individual variability. The superiority of the average I.Q. among the children of the well-to-do as compared to those of the poorer classes, is an example of the dual set of causes. Psychology is so intimately biological that its principles and practices depend upon the issues of patterns more definitely traceable in bodily traits.

In further illustration of the controversial character of issues in psychology which must find their illumination under biological principles, there is the eternal problem of the eternal feminine—the correct view of the differences in mental organization of men and women. For here are three vital determinations of original nature which shape the fate of humanity and the course of civilization—the race, the sex, the individual or familial stock.

The racial question is too complex for summary inclusion.[1] The order, the extent, the interrelation of racial traits are all written somehow in the varieties of nervous organization: they are determinations of nature and cumulate with those of sex and stock. We reach the individual through common remote (racial) and immediate selective ancestry, everywhere differentiated by the (glandular) determinations of sex. Racial intermingling, as well as the large ancestral community of all races, served to obliterate differences. But such an experiment as transporting West African negroes to a wholly different environment still leaves intact the racial differentiations, as shown by the fact that the average I.Q. of the negroes drafted in the war was 10.4 years, while that of the mixed whites was 13.6 years. The capacity of races to adopt and adapt themselves to alien cultural environments is demonstrated by the same experiment.

The question of interpreting biologically the differential psychology of men and women comes back to that of secondary or derivative sexual characteristics. The physical differentia admit of no dispute; nor are we under the temptation of confusing conspicuousness with importance. That men have beards and women have none gives us no concern; if nature wills it that way, we accept the dispensation or nullify it with a razor. The functional differences we accept with equal calm, noting the different distinctions, such as liability to disease, or the varying emotional nature centering about the care of the child. But the intellectual sphere under the dominion of the highest orders of cortical organization presents a different problem, which has disturbed empires, created political entanglements, and turned economists, sociologists, psychologists,

[1] For an example of its treatment under the guiding principles of modern psychology, see S. D. PORTEUS and M. E. BABCOCK, "Temperament and Race," Richard G. Badger, Publisher, Boston, 1926.

[88]

philosophers, and writers generally into propagandists,[1] particularly in the intimate issue of the relation of the sexes which is pivotal to the quality of social life.

Achievement is an uncertain test of races for one reason, of the sexes for another; yet they are related. The social environment can go far to disqualify the status of women. The conforming power of social pressure is enormous; that alone is a testimony to the large truth of the environmental psychology. If one were to judge the relative intellectual ability of the two sexes by the frequency of their citation in a standard biographical dictionary, one would reach a conclusion contradicted by all biological evidence. Whatever the scale of difference between the masculine and the feminine intellect, it is not of that order. Opportunity, stimulation, expectation, supporting qualities, the place of career in the total scheme of life—social disqualification as well as original nature—all play a part in the disparity and leave an insoluble equation, because of too many unknown quantities.

But when the experimentalist compares the school ratings or college standings in a range of artificially selected abstract studies or in elections to Phi Beta Kappa, and concludes, because these are so nearly equalized in young men and young women students, that the commonly accepted views of the mental differences between men and women are the result of prejudices, hearsay, myth, and convention, he is greatly overestimating the value of his technique and as greatly neglecting the biological scale by which all such values must be gauged.

These scholastic tests are adventitious and tangential in any biological scale; they are not likely to touch the significant differences that support such remote applications

[1] Sufficiently illustrated in the volume by various authors, edited by V. F. Calverton and S. D. Schmalhausen, "Sex and Civilization," Macaulay Company, New York, 1929.

of endowments. And even there it is far more significant that such selection as diminishes the proportion of women students in law and engineering to nearly zero, as determines what branches within the academic program attract women and in which they excel; that women in general are as musically inclined as men if not more so, yet few women become composers; that the stage arts in which personality counts, are the ones in which women hold their own; that by far fewer women carry their proficiencies (writers excepted, again in the field in which personalized human relations count) to a professional status—all this is certainly of far larger consequence than a set of proficiencies which happen to be more measurable by the aid of assumptions. The confusion of measurability with importance is the experimentalist's fallacy. It can be corrected only by loyalty to the scale—if we can but learn to read it—of nature's significance. The arguments thus suggested pro and con are fragmentary; they are sufficient to indicate an approach to the solution of the problem involved by way of a naturalistic appraisal. It will then be found that men and women are even more profoundly and variously different than we have yet discovered. How could it be otherwise, since the principle of physiological psychology is that bodily functions are reflected in psychic ones in all degrees of radiation; and there is no deeper and more pervasive physiological difference than that which divides human beings (abstractions) into men and women (as realities). Confusion results from neglect of this distinction. Such a principle as the Adlerian "masculine protest" is left floating in mid-air when applied to women. In the future we may recognize a masculine and a feminine psychology. The practical bearings of it all upon adapting the social and industrial arrangements of the world to the needs of men and women are vast. Circumstance will continue to shape institutions. But there is no problem that cannot be approached in the spirit of scientific inquiry, even that of the desirable relations between men and women.

The note thus touched upon is important. The larger fields of application of modern psychology are those of political, industrial, national, and international concern. There is a psychological approach to almost every public question. To embark upon such an experiment as nation-wide prohibition without the benefit of expert psychological examination is but a flagrant instance of the unwisdom invited by democracy tempted to the ways of demagogy. The fact that the alcoholic temptation lies so close to the crime issue, that it represents a psycho-neural craving as old as mankind and as widespread, that it holds a close relation to the masculine psychology of escape and search for intensive thrills—all this should have indicated the province, along with the allied specialties of physiology and psychiatry, in which the policy of wisdom in regulation was to be sought. But political eyes are blind, and political ears are deaf to the appeal of calm scientific investigation, when issues must be reduced to votes and the ordinary limited comprehension and prejudices of voters.

Yet prohibition is a small concern, despite its wealth of lessons in wrong ways of righting ills, as compared with the stupendous dimensions of war, which is likewise a psycho-political issue. To recognize that man has ever been his own worst enemy, that the most intelligent of species is the only one that engages in collective extermination on a race-suicidal scale, is as far as the reference can here be carried. It is germane for its inclusion of the mass psychology in which social psychology finds its culminating problem and expression. Human gregariousness is a formidable quality; it makes the cosmopolitan centers, crowding humanity into congested quarters, conferring the benefits of cooperative enterprise, and creating the larger loyalties that make patriotism possible and national ideals realizable. Yet it sets nation against nation and multiplies to colossal proportions the menace of conflicts inherent in the contentious motives of the industrial members of the human herd. Collective psychology is itself a specialty, considering the contagion of example, the spread of prejudice, the vogue of fad and fashion, the lowered mentality under crowd pressure, the insanity of the frenzied mob. To reconcile the social with the industrial nature, to direct social pressure wisely and form public opinion judiciously, is the problem of civilization. The Freudian recognition of the conflict is in the name of the censor, which may be a Puritan conscience or a worldly Mrs. Grundy, a Calvinistic restraint, or an aristocratic oppression. Through it all we recognize that man cannot follow his original nature and become a useful citizen or a civilized being. Man is the only animal who reconstructs his own nature, and in that process finds the glorious opportunities of realizing his potentialities. Civilization is psychological reconstruction.

Psychology makes no pretense to be all things to all men. It aims to be a naturalistic science among the sciences, but deals with the psychic realm which is alike the instrument of human understanding however directed, and also the subject matter, when the mind explores its own nature. The renaissance of psychology is a story of rapid expansion within hardly more than a single generation. It seems worth the effort to include in a single canvas, however crowded, the movements of ideas and inquiry contributing to its recent advance, and to do so from the unifying point of view of a naturalistic psycho-biology.

Psychology, in coming in line with the natural sciences, plays its part in the rationalization of knowledge, and an important one. Many of its fundamental problems are ancient; but the solutions shift with the temper of the times, with the advancement of learning. There is an evolutionary psychology in sequence of development. The spiritistic and the moralistic trends dominate in primitive psychology. Man formulated his "soul" in those terms. His purpose was to appease the gods and control his own fate. His psychology was shaped in the image of his desires.

As learning became established, the doctrines became academic and solutions traditional. Authority and consensus established what was held to be the truth and ever in fair conformity to religious, moral, or social codes supported by established authority. Because of the lack of knowledge of nature's ways, a pretentious and assertive pseudo-science flourished, mingling with folklore beliefs. Parallel to the progression from alchemy to chemistry, from astrology to astronomy, is a similar growth from a primitive, a religious, and an academic psychology to a body of doctrine that is conceived in the spirit of natural science and proved by way of experiment.

The conclusion to be drawn from the renaissance and its antecedents is that issues once regarded as lying outside the realm of the methods of science may be pursued in the same temper and logic as have brought about the great advances in the physical sciences and the control of natural resources. This does not mean that psychology will be reducible to physics or physiology; its distinctive province remains and will ever require adaptation of technique to the conditions under which the phenomena occur. But it has become possible to focus psychological findings upon the conduct of human affairs, to recognize the voice of psychology in many concerns formerly decided by the guidance of religious precept or legal enactment, and above all to naturalize psychology in the federation of the sciences. Upon the wise application of the principles thus emerging depends in no small measure the direction of future progress.

[93]

Chapter IV

EDUCATIONAL PSYCHOLOGY

by Lewis M. Terman

PSYCHOLOGY as a scientific discipline is only two generations old, the first laboratory having been founded in 1879. Educational psychology is considerably younger, for it was only natural that the early workers in the new field should be completely engrossed in devising investigational methods and in laying the broad outlines of the science as a whole with little regard to its practical values. When later the importance of psychology for education began to be recognized, treatises were written which explained the applications of psychological laws to the teaching process. For some time educational psychology was thus nourished almost entirely from crumbs that fell from the table of the so-called "pure" psychologist, with the result that the latter was often inclined to regard it with a certain amount of contempt. When later the problems of educational psychology began to be investigated on their own account, it was found that the line between pure and applied science is more imaginary than real and that science itself is largely dependent for its growth on the motivation that has its source in practical issues.

To-day educational psychology is no longer *applied* psychology in the old sense of the term. It is one of the most important aspects of the general science, just as psychopathology is, and just as animal behavior is. Its methods, devised for such specific purposes as investigating indi-

[94]

vidual differences, mental development, and the facts of learning, have profoundly influenced psychological methods in general. Its results, likewise, have been incorporated into the science and are helping to shape psychological theory. The educational psychologist is merely a psychologist who prefers to work with problems that happen to have a bearing upon education; actually his personal interest may have less to do with education, as such, than with the eternal verities of human nature.

If we define education as the making of changes in human nature, then nothing which concerns human nature can be entirely alien to it. One might go further and say that nothing which concerns the behavior of any biological organism can be entirely alien to it, for there are important principles that apply to all living things. Nevertheless, there are certain aspects of biological science, and more particularly of psychological science, that concern education much more than others, and it is these only that can be touched upon in this cursory review. The present discussion will therefore be limited to the following topics: (1) instinctive tendencies, (2) individual differences and their causes, (3) psychometric methods, (4) mental development, and (5) the psychology of learning.

Instinctive Tendencies. If the purpose of education is to make human behavior different from what it would be without education, it is necessary to take account of what original nature has given. This must be done in two directions: (1) we must consider instinctive tendencies, and (2) we must take account of capacity to modify original nature—that is, of intelligence or ability to learn. In both of these types of endowment there may exist individual differences over and above the traits which characterize human beings as such, and in the field of intelligence the differences are especially important. We shall consider first

[95]

the problem of human instincts, one of the most controversial issues in present-day psychology.

The point of view made popular by William James, that
man has many well-defined instincts—more than other
animals, rather than fewer—was generally accepted by
educational psychologists for a quarter of a century.
The long list of human instincts compiled by James was
followed in 1906 by the similar though somewhat briefer
list of McDougall. McDougall's presentation of the instinct problem, which for a decade or more greatly influenced thinking in educational as well as in social
psychology, had much in common with that of James.
Both regarded instincts as primary determiners of a great
fraction of human behavior, as psychobiological forces
which give direction to all our striving. According to
McDougall the essential nature of an instinct is not to be
sought in the exact mode of response to a particular
stimulus, but in its goal-seeking and purposive character
which may cause it to find expression now in one type of
response and now in another. Then in 1913 followed
Thorndike's treatment of the subject, somewhat more
in the vein of James than of McDougall, but resembling
both in the extended list of instincts posited. For Thorndike, however, an instinct is an innate bond between a
specific stimulus and a specific response, and not the
generalized psychophysical disposition assumed by Mc-
Dougall. Instincts in McDougall's sense he regards as
pseudo-entities that have no existence in reality. The
important thing to know is the particular response that
follows or tends to follow a particular stimulus. Thorndike's psychology of instinct, like his psychology of learning and of intelligence, is a *bond* psychology. It is thus
essentially a revamping into biological terminology of
the older *association* psychology given vogue by Locke,
Hume, the two Mills, and Herbart.

In recent years there has been a strong, not to say violent, reaction against the older doctrines of instinct, especially against that of McDougall. This reaction has taken the form of gross reduction in the number of instincts posited, and has even gone to the extreme of denying that any instincts, either human or animal, exist at all. The denial of instincts to animals need not, of course, be taken seriously. To believe that all the myriad behavior patterns of animals are learned rather than innate is obviously absurd. The arguments against the existence of human instincts, however, are more plausible. If we follow the behaviorists in defining instinct as a specific response to a given stimulus, then it must be admitted that, with the exception of a small number of reflex responses in the young infant, there are no human instincts. Apart from these, the child's behavior is soon modified to such an extent by learning that few well-defined, persistent, and universal pattern reactions are discoverable.

According to the extreme behaviorists, practically all of a child's reactions are the result of "conditioning," in Pavlov's sense. Human beings *seem* to have instincts of the sort designated by such terms as pugnacity, fear, sex, maternal love, gregariousness, ascendency, submission, etc., only because of the conditioning effects of social environment. It is argued that, given appropriate environmental stimuli, a child could be caused to grow up without evincing any of the modes of behavior we are accustomed to designate as instinctive. From this point of view, the newborn infant, if not exactly a blank tablet, is at most a wriggling mass of random activity plus a few minor reflexes. On this almost formless foundation, it is argued, any kind of structure can be raised by the simple method of conditioning. Watson promises to take any normal healthy infant and to mould him to any pattern whatever—that of musician, artist, scientist, fool, hermit, social lion, introvert, extravert, coward, hero, lover, hater of sex, or anything else. Education counts for everything, original equipment for nothing. According to this view, if educational psychology is to make any progress, it must give up its superstitious belief in instincts and learn how to build up the behavior patterns it wants by the method of the conditioned response.

Actual data in support of so extreme a view are very scanty. The few experiments that have been made with infants do show that, in the field of the emotions, conditioning plays an important rôle. Whether a child's fear responses are called forth by snakes, furry animals, horses, dogs, mice, automobiles, trains, or by none of these agents, depends upon experience. It is a far cry from such facts, however, to the conclusion that the possibilities of conditioning are unlimited and that it takes place in one direction as readily as in another. Such a conclusion is belied by the fact of the essential similarity of human nature that is found in all the widely varying types of social environment. For example, although the *mores* that have to do with courtship and mating vary infinitely, the part played by sex in human relationships is very much the same the world over. The same is true of pugnacity, fear, maternal love, jealousy, gregariousness, ascendency, submission, etc. The overt behavior in which these native tendencies are expressed takes on every conceivable form, but the motivating influences are everywhere the same. The native equipment of the human being predetermines him to differ in important respects from any other animal, even from the anthropoids which are biologically nearest to him. No amount of conditioning is capable of making a Goethe out of a gorilla. To make of the average normal human infant an unselfish saint, utterly unconcerned about his food, his sex needs, and his relations to other human beings is difficult enough. The natural tendencies to anger, pugnacity, love, fear, jealously, gregariousness, and other forms of behavior can only be overcome by an extraordinary amount of training, and then only partially. To the extent that they can ever be overcome, this is only possible by the utilization of antagonistic or substitute tendencies that are equally native.

From human nature there is no escape for anyone. The psychologist who would lay down canons for the education of human beings can no more safely leave native tendencies out of account than the trainer of dogs, fleas, elephants, lions, and seals can ignore the original behavior tendencies of these animals. The limitations of conditioning are seen in the fact that no one has ever succeeded in training an animal of one species to behave like an animal of another species. The gregariousness of monkeys, the solitariness of eagles, the wildness of the zebra, the pugnacity of the male seal toward other males, the social behavior of the bees, the feeding activities of the owl, and the maternal behavior peculiar to every species of birds and mammals, are tendencies that can not be ignored by one who has anything to do with these forms of animal life. That human beings are an exception to the rule with respect to the actuality of psychophysical predispositions is wholly unreasonable.

The fact that in human beings the appearance of native tendencies is often delayed until much learning has taken place, in the case of some of them until adolescence or later, means that instinctive drives cannot be expected to show themselves in exactly the same overt behavior in all. That is to say, there are in human beings no *pure* instincts, but only instinctive tendencies mixed with habit in varying proportions. Perhaps it would be well if the term *instinct* were dropped altogether from human psychology in favor of some such term as *instinct-habits*, or *coenotropes*, as they are called by Smith and Guthrie. I would suggest the term X-Y traits, indicating that the behavior trait, as we find it, is a compound composed of two elements combined in unknown proportions—X representing original nature, and Y representing the factor of nurture.

The scientifically minded psychologist naturally regrets that the relative contributions of these two factors cannot at present be precisely delimited; but for the practical purposes of education this is not so important. What is important for the educator is to know what X-Y traits are present in children when they present themselves in the schools, and how best these traits can be used in the process of moulding children according to the patterns furnished by the social ideals. Just what traits shall be included in such an X-Y list will depend upon where the line is to be drawn with respect to the amount of saturation with the X element before the trait is so classed. In one sense, every trait the child exhibits is an X-Y trait. It always contains some element of original nature and some elements that have been contributed by experience. Limitation of the number to one, such as the Freudian sex motif or Adler's will-to-mastery, ignores a large part of the total realm of human behavior. To extend the list to include only anger, fear, and love, as does Watson, also fails to do justice to the richness of human nature. I forbear to add a list of my own, but would refer the reader to the lists of James, McDougall, and Thorndike as representing what I regard as, on the whole, reasonably satisfactory inventories.

In order to deal wisely with human nature in its formative period, the educator must take account of the natural tendencies of human nature to fear, anger, pugnacity, disgust, curiosity, self-assertion and self-display, self-abasement, sex love, sympathy, tenderness, gregariousness, imitativeness, suggestibility, play, jealousy, and other types of behavior. There is no need to decide in the case

[100]

of each of these traits how much is original and how much is acquired; it is necessary to take account of all of them as persistent motives.

There are certain types of fears which children are prone to develop and definite ways in which the fear motif in their lives can be reduced or directed. There are many types of experience which tend to provoke anger; and in some the anger response is to be encouraged, in others discouraged. If curiosity in the form of natural childish interests is neglected, the result is likely to be the imposition of educational subject matter and procedures which run counter to child nature. Neglect of the deep-seated desire for self-assertion and mastery brings about the formation of inferiority complexes which may paralyze activity or result in over-compensation in the form of conceit, "touchiness," and boastful lying. Neglecting to take account of gregariousness, suggestibility, and imitativeness creates half the problems of school discipline, gives us most of our juvenile delinquents, and renders ineffective some of the best-intentioned efforts of moral education as it is practiced. To ignore the appeal of play and make-believe in the life of children would place us below the most primitive savages in educational wisdom. As for the pervading influence of sex, so much has been written on this subject in recent years that its importance is now generally recognized. It is possible here merely to call attention to some of the X-Y traits which the educational psychologist must take into account; to attempt to formulate desirable educational procedures in connection with each would require a volume.

Individual Differences and Their Causes. One of the most important of the contributions to education which psychology has made is in the field of individual differences. The older psychology was concerned almost entirely with the establishment of general truths concerning human beings; modern psychology is no less interested in the ways in which human beings differ. As soon as attention was focused upon individual differences, it was found that these were omnipresent and very large. It was found, for example, that people differ far more in intellectual ability than they do in height or weight or other physical traits, and that the intellectual differences among school children are a more important factor than pedagogical methods in determining how much a given child will learn. Later, experimentation was begun in the measurement of special abilities and of personality traits, with the result that large individual differences were found wherever a search was made.

One of the leading characteristics of present-day education is the almost universal effort that is being made to shape the work of the school to meet the individual needs of its pupils. Such attempts are affecting not only the pedagogical methods employed, but also goals of instruction, the control of curriculum, and the educational and vocational guidance of children. Special classes are being established for the gifted, the backward, and the predelinquent. Problem children of all types are psychologically studied, and differential treatment is provided not only in the schools but also in the courts and the reformatories. Earlier types of impressionistic case studies are being largely replaced by psychometric techniques which yield quantitative measurements of general intelligence, of special ability in art, music, science, and mathematics, and of such personality traits as ascendency-submission tendencies, inferiority feelings, introversion-extraversion attitudes, mental masculinity and femininity, fairmindedness, social intelligence, emotional stability, trustworthiness, and occupational and other interests. In all such traits it is found that the range of scores made by unselected children is about as great as it is for intelligence. There is

[102]

still an immense amount of work to be done in devising new instruments for psychological measurements, improving old ones, and finding the relationships which obtain among various combinations of traits. Nevertheless, the psychology of personality, which only a few years ago seemed too elusive to permit a scientific approach, is rapidly yielding to quantitative treatment. Even temperament and the psychological correlates of physical types are now investigated by scientific methods.

In the use of psychometric techniques it is not necessary to start with any assumption in regard to the cause of the individual differences found. Their purpose is to measure the differences which exist, regardless of how they may have originated. However, when once a satisfactory technique for the measurement of a particular trait has been perfected, it is then possible to investigate the relative contributions of nature and nurture influences. This has been done most extensively in the case of intelligence, but to a less extent also in the case of certain special abilities and personality traits. The existing data pertaining to the relative influence of nature and nuture have been judiciously summarized in Part I of the 1928 *Yearbook* of The National Society for the Study of Education. The issue is an extremely controversial one, and it is impossible to review here the factual evidence in detail. My own interpretation of the data will be stated in a few brief paragraphs.

In the case of intelligence, as measured by the best intelligence tests, I consider the evidence overwhelming that the larger differences found among children in an unselected school population of a typical American community must be credited to nature rather than to nurture factors. The mere fact that the correlations between parents and offspring and those between siblings are almost always between .40 and .50 in intelligence test scores, of course, proves nothing either way. The correlations can be accounted for on either the nature or the nurture hypothesis. Facts which seem to me to render the nurture hypothesis untenable are the following:

1. Although twins of two-egg origin resemble each other in intelligence no more than do ordinary siblings, twins of one-egg origin (having presumably identical heredity) are in the vast majority of cases of almost identical intellectual ability. In fact no very large differences have ever been reported for identical twins who have been reared apart.

2. Foster children adopted in the first year of life have been shown by Burks to yield almost a zero correlation with their foster parents in Stanford-Binet intelligence scores. Burks' data indicate that the contribution of heredity to variational differences among children of modern American communities is about five times as great as the contribution of the nurture factors connected with the home. It is true that Freeman's study of foster children was considerably more favorable to environmental influences, but the children he studied were in most cases adopted at later ages, when selective influences could have entered to distort results. Obviously the only way to avoid the selective factor in such a case is to study only foster children who were adopted so early in life that it would not be possible for the more intelligent parents to select the more intelligent children.

3. Children of widely different grades of intelligence are found in exactly the same environment. Siblings, for example, are found to differ over a range from an intelligence quotient of 50 or less to one of 150 or more.

4. Individual differences in intelligence are evidenced from very early years. They are readily detectable by the age of twelve months and at thirty-six months are about as marked as they are at any later age. One would certainly not expect this to be the case if nurture factors were paramount.

[104]

5. Experimental attempts to improve the intelligence of backward and defective children have met with extremely little success. If I knew how to raise the true I.Q.'s of ordinary public school children by 25 points, say from the level of 75 I.Q. (extreme dullness) to 100 I.Q. (average intelligence), I could be a millionaire within ten years. I do not know how, nor does anyone. One can increase the found I.Q. in a given intelligence to this extent by specific coaching on the test questions, but this is a spurious improvement; on other tests and in learning ability the child remains practically what he was before.

6. Finally, the fact that white rats show marked family resemblances in ability to learn mazes and to solve other problems requiring intelligence argues for the hereditary nature of intellectual differences generally. The actuality of family resemblances in the case of rats has been decisively demonstrated both at Stanford University and the University of California. These resemblances certainly cannot be explained as due to cultural influences!

The inheritance of special abilities has been less extensively investigated for the reason that not many kinds of special ability have thus far been successfully isolated and measured. By means of the Seashore tests it has been demonstrated that the elemental factors of musical ability show a very marked tendency to run in families and to be but slightly improvable by tuition. There is much genealogical evidence that the same is probably true of artistic and scientific ability, but owing to the difficulty of ruling out nurture factors the evidence from this source is not conclusive. A child's mechanical ability has been found not to be greatly influenced, as a rule, by the mechanical environment in the home.

About the inheritance of character and personality traits still less is known. Dr. May's tests of trustworthiness have shown nearly as great resemblance between siblings in proneness to cheat, lie, and deceive as others have found for intelligence; but the possibility must be freely admitted that such correlations may be explained as merely reflecting the moral atmosphere of the home. However, F. A. Woods finds that in the royal families which he studied, moral traits characteristic of a certain stock often appear very persistently among branches of the family that have been entirely separated from each other for a generation or more. There is evidence enough that tendencies to psychopathic disorders are transmitted by inheritance. It is probable that important differences in temperament are innately determined, perhaps as correlates of inherited glandular traits. Kretchmer and others have attempted, not entirely without success, to relate certain personality types to types of body-build. Herrington has found reliable, though fairly low, correlations between introversion-extraversion test scores and various chemical reactions of the body; and Rich has found chemical correlates of emotional excitability. No one doubts that the easily recognizable temperamental differences among various breeds of dogs, horses, and other domestic animals are, for the most part, innate.

When we come to such personality traits as fair-mindedness, radicalism, conservatism, or mental masculinity and femininity, the nurture factors loom much larger. These traits are the product of a great variety of attitudes and interests which seem to be readily moulded in early life. The study of mental masculinity and femininity, by means of a test specially devised for the purpose, shows (1) that one's rating in this trait bears almost no relation to masculinity or femininity of body build; (2) that family resemblances in the trait are low; and (3) that extreme inversion in the trait can often be accounted for by environmental factors. Omitting from consideration cases of complete or partial hermaphroditism, which are extremely rare, it appears that the vast majority of persons with inverted sexual inclinations (that is, homosexuals) are made so by their early experiences. Parent fixations are common causal factors. As from 2 to 5 per cent of the general adult population show more or less marked homosexuality, either overt or latent, it is evident that the problem is much more serious than it is usually supposed to be.

Generally speaking, it is in the cultivation of desirable traits of personality that education can exert its greatest influence upon original nature. The best of training over and above that which is now commonly available in the schools is not capable of adding so very much to a child's intellectual stature. By suitable methods of instruction the dull may be made to learn a little more than they otherwise would, and by the choice of suitable curriculum material they can be equipped for life far better than they usually are, but their ability to carry on complicated intellectual processes remains practically unchanged. And just as dull children are not made bright by good teachers, so bright children are not made dull by poor teachers. In the case of personality traits the situation is very different. There is reason to believe that it is possible to make a child deceitful or truthful, honest or dishonest, fearless or fear-ridden, fair-minded or prejudiced, religious or non-religious, masculine-minded or feminine-minded, almost at will. It is unfortunate that education neglects most the very things it can do best.

Psychometric Methods. Since individual differences are so important for education, and since they cannot properly be taken into account without quantitative methods of measuring them, it is desirable to give the reader a more definite idea of the technique which such measurements employ. There are now hundreds of psychological measuring instruments, commonly known as mental tests, and it is manifestly impossible to describe them all in this brief treatment of the subject. Fortunately most mental tests have certain important principles in common, and these can perhaps be made clear by giving first an account of the methods of measuring intelligence.

When one thinks of the measurement of intelligence, one inevitably thinks of Binet, the great French psychologist who labored so diligently in this field for more than twenty years (1890 to 1911). It is Binet more than anyone else to whom we are indebted for modern methods of intelligence testing. It is true that Galton had preceded him by fifteen years in appreciating the importance of individual differences and in the first crude attempts to measure them by means of tests; also that Kraepelin in Germany and Cattell in America antedated Binet in their entrance into this field. It was the genius of Binet, however, that most clearly envisaged the problem and that gave us the first workable methods of testing intelligence.

The early tests had confined themselves chiefly to the simpler mental functions, such as sensory discrimination, perception, motor coordination, reaction time, and rote memory. Large individual differences were discovered in all these fields, but they were not found to be correlated in any great degree with intelligence in the ordinary sense of that term. Binet, after many explorative researches in which he investigated the relation of these and other traits to intelligence, abandoned tests of the simple mental functions in favor of tests of such complex functions as logical memory, controlled association, comparison of concepts, detection of absurdities, comprehension of difficult situations, definition of abstract terms, problem solving, etc. Such abilities, Binet found, were highly correlated with ability to master the school subjects and with social efficiency. It is in such traits that the recognizably feeble-minded differ most from normal persons, not in the simple sensory, motor, and perceptual functions.

[108]

Intelligence testing is nothing more nor less than a sampling of one's intellectual and behavior responses. Galton likened it to the assaying of mineral ores. The effectiveness of the sampling depends upon: (1) whether it is extensive enough to give consistent results, and (2) whether it is a sampling of intelligence or of something else. If a test is extensive enough to give consistent results, it is said to be *reliable*. If its verdict correlates fairly well with other criteria of intelligence, the test is said to be *valid*.

The reliability and validity of tests cannot be judged by opinion, either the opinion of the layman or of the psychologist. Both have to be experimentally determined. The results of such determinations are expressed in quantitative terms, usually correlation coefficients. It has been definitely established that the best of the current intelligence tests are reliable in the sense that they give consistent results, and that they are valid in the sense that they discriminate clearly between groups of subjects classified by other reasonable criteria as bright, average, or dull. Whatever misinformed critics may say, the best intelligence tests really do measure with greater or less approximation what we may reasonably call intelligence; that is, ability to make complex adaptations, to think in conceptual terms, and to solve difficult problems.

The limitations of intelligence tests must, however, be kept in mind. These are chiefly of three sorts. In the first place, no intelligence test is a test directly of native ability. It tells us what grade of intelligence the subject has when he is tested, not necessarily what he might have had if his education and cultural background had been different. Any judgment about native intelligence is always a matter of inference. My personal judgment is that the experimental investigations to date do not suggest that the scores on the best intelligence tests are very greatly influenced by ordinary physical defects (such as adenoids or malnutrition) or by moderate differences in regularity of school attendance, in pedagogical skill of teachers, or in the cultural level of parents. At the same time it is freely admitted that extreme differences in regard to such advantages and disadvantages may and probably do, in some degree, invalidate the test result. Ultimately the influence of these and other possible determinants of intelligence will be measured by the test method. In the second place, one must bear in mind that even the best tests are not perfectly consistent, or reliable. Every test score has its probable error. This has been carefully ascertained for all of the better types of tests. In the third place, one must remember that no test samples every aspect of intelligence; one may emphasize ability to deal with abstractions, another the ability to adapt to common sense situations, and another aptitude for working with things.

A subject's test performances have little meaning until they have been compared with norms. The norms that have proved most useful are those based upon age. A subject whose score equals that of an average six-year-old is said to have a mental age level of six years; if the score equals that of the average ten-year-old, the mental age is ten years, etc. A particular subject whose actual chronological age is eight years may have a mental age of four, six, eight, ten, twelve, or even fourteen years. Dividing the mental age by the chronological age gives us an index of brightness known as the intelligence quotient, or I.Q. For example, the eight-year-old of six-year mental age has an I.Q. of 75; the eight-year-old of ten-year mental age, an I.Q. of 125.

The educational significance of mental age lies in the fact that it determines what a subject is able to learn, what can be profitably taught him. Children entering school with a mental age of six years can be taught to

[110]

read; when the mental age is much less than this, their reading progress is slow. On the other hand, a child who enters school at the age of six or seven but with a mental age of nine or ten years is able, if given a chance, to cover all the work of the first three grades in the first school year.

The educational and social significance of the I.Q. lies in the fact that for a given subject it tends to be fairly constant throughout the period of growth. The child whose I.Q. is 75 at the age of six is not found with an I.Q. of 125 or 150 a few years later. Increases, decreases, and fluctuations of I.Q. do occur, but in the large majority of cases they are rather small in amount, rarely more than 15 or 20 points, and on the average not more than 5 or 6 points in a total range among school children from I.Q.'s below 50 to I.Q.'s above 150.

The child below 60 I.Q. is ordinarily classifiable as feeble-minded and is not likely ever to develop beyond the mental age of eight or nine years, as intellectual growth pretty well ceases by the age of fifteen or sixteen years. Subjects in the range between 60 and 80 I.Q. may or may not be able to make their way in the world, depending largely on their emotional, volitional, and other non-intellectual equipment. Those of 75 to 90 I.Q. may be classified as normal but dull, those of 90 to 110 I.Q., as average normal, those of 110 to 130 as superior, and those above 130 as very superior. About ten children in a thousand test as low as 70, and about an equal number as high as 130. On the average four or five in a thousand are above 140 I.Q. About 15 per cent test below 80 I.Q. and about an equal number above 120 I.Q. It is from those above 115 or 120 I.Q. that we get the majority of our college students. Those as low as 90 are rarely able to graduate from high school, and those as low as 80 rarely progress beyond the seventh or eighth grade.

Intelligence tests are widely used in the schools in grouping children for instruction. Teaching can never reach optimum efficiency in classes of extreme heterogeneity. The class range of ability is reduced by division into *A* and *B* sections, and still more by a three-fold or five-fold classification. Classification of children into *X*, *Y*, and *Z* groups (corresponding to bright, average, and dull), with special classes for the feeble-minded and for the most highly gifted, has been found to yield the best results. *X* and *Z* groups, for example, do not profit maximally from the same type of curriculum or from the same pedagogical procedures. The methods and subject matter of special classes for defectives have little in common with those suited to average children, and much the same can be said of special classes for the highly gifted. The widespread introduction of special classes for the gifted is unquestionably one of the most significant movements in modern education.

Classification of a given child for instruction is not based upon intelligence tests alone, but takes account of health, previous scholarship, and personality factors. Nor is placement looked upon as final; the progress of each child is followed, and classification is changed when it appears desirable. The aim is to fit the school to the child in order that each may profit to the greatest possible extent from the instruction given. This makes it necessary to have objective measurements, not only of intelligence, but also of as many traits of mind and personality as it is possible to secure. The result is that psychologists are busily engaged in devising, validating, and standardizing scales for the measurement of achievement in all the school subjects, others for the measurement of special aptitudes, and still others for various aspects of personality, interests, and attitudes. In most cases the tests of personality traits must be regarded as tentative and experimental, but they are opening important fields of research and will in time give a much more scientific basis for education than we now have.

The most successful tests, other than those designed to measure general intelligence, are the standardized tests of achievement in the school subjects. These are in part replacing the old-fashioned school examinations which were so notoriously unreliable and subjective. The new-type examinations (the standardized test) call for a far larger and fairer sampling of the child's knowledge and yield scores which are not influenced by the bias of the examiner. When they are used in connection with intelligence tests, the promotion and classification of pupils is effectively removed from the realm of guesswork.

The tests of musical ability help to identify, even in the lower school grades, the children who have musical promise. In the past, millions of dollars have been expended annually on the musical education of children who were almost wholly lacking in musical talent, while countless others of superior natural talent remained undiscovered. The same is true with respect to artistic talent, mechanical aptitude, and other types of special ability. The rapid progress which is being made in psychological testing warrants the belief that in the not very distant future the vast majority of school children will be measured for several kinds of special ability and special defects as well as for general intelligence and for previous scholastic achievement.

The personality tests will doubtless have less effect on ordinary classroom procedures, but their ultimate influence upon child training in the broader sense is likely to be profound. It is already apparent that the early treatment of children by parents is capable of warping their personality traits for life in definitely measurable ways. Lack of opportunity for self-expression breeds inferiority complexes

[113]

that interfere with normal personality development. Lack of normal social contacts, especially if this is combined with a sense of inferiority, leads to extreme introversion. Too much mothering of the boy or too much fathering of the girl warps the child in the direction of homosexuality. Attitudes of honesty, fair-mindedness, and tolerance in home and school definitely affect the scores which children make on tests of these traits. Such influences have long been suspected; they are now being quantitatively measured. In the case of honesty and trustworthiness, for example, it has been experimentally demonstrated that the child is much less influenced by what he is taught verbally than he is by the general atmosphere of honesty and truthfulness in the classroom or home.

The measurement of interests is especially promising as an aid to vocational guidance. It has been shown that the child himself cannot be depended upon to give a correct account of his interests. He easily misjudges his interest in a given school subject because of his attitude toward the teacher; he misjudges his interest in a given occupation because of his admiration for some one in that occupation. In fact the child has no possible means of knowing whether his interests and attitudes most resemble those of lawyers, doctors, engineers, artists, teachers, business managers, or some other occupational group; but interest tests have been devised to answer just this question. They cannot serve as the sole basis of vocational guidance, but they promise to become indispensable aids.

The above are only examples of many illustrations that could be given to show how the study of individual differences is influencing education. In the mass education so necessary in a democracy, individuality has always been sacrificed in greater or less degree in favor of a uniform product. But a democracy, to survive and progress, demands not only a high level of general attainment in its population, but leadership as well. Psychological measurements of the kind we have described are making it possible to identify exceptional talent of every kind and to rescue the individual from the mass. Many problems remain to be solved before this end can be accomplished, but substantial progress is being made.

Mental Development. Since the beginning of the child study movement in the eighties of the last century, innumerable researches have been published in the psychology of mental development. These have included psychobiological studies of various life periods, especially of early infancy and adolescence, studies of language development, growth in character and personality traits, and age changes in general intelligence, special abilities, and interests. The amount of such research is constantly increasing, and its character is becoming more scientific. The child-study enthusiast of forty years ago was usually ill trained in scientific method, and his results were of little or no value. At present there are in the public schools, in institutes for child welfare research, in child guidance clinics, in juvenile courts, and in institutions for defectives and delinquents hundreds of well-trained psychologists investigating the problems of mental development. It is possible here only to indicate the types of problems which are being studied and to state briefly a few of the more important conclusions to which investigators have been led.

The rate of mental growth was formerly thought to be irregular and to give the effect of well-defined stages. It has been found, on the other hand, that development is gradual and regular, almost totally lacking in saltatory character. Early childhood is not preeminently a sensory stage, middle childhood a motor activity stage, pre-adolescence a memorizing stage, and adolescence a stage of rapid intellectual growth. It is true that any two periods,

[115]

separated by a few years, present fairly clear differences, but new capacities rarely, if ever, develop suddenly. Generally speaking, mental growth proceeds at a gradually decreasing rate from infancy to maturity. It is greatest in the first year, somewhat less in the second year, still less in the third, and so on, until by the age of fifteen or sixteen the changes from year to year have become relatively small. It is possible that interests and attitudes are subject to more sudden changes than are intellectual abilities, but for the most part they too run an even course. Even the rebirth which adolescence is popularly supposed to bring is largely a myth. Still less is there ground for believing, as Stanley Hall did, that the child in the period from eight to twelve years is especially adept at memorizing and incapable of any genuine reasoning, or that he lacks appreciation of moral values and responds only to authority. Sweeping and untrue or half-true generalizations of this kind have done much harm, especially in justifying radical shifts in educational procedures from grade to grade. Until very recently, for example, kindergarten methods used with five-year-olds differed so completely from the methods used with six-year-olds in the first-grade that one would have supposed children of these two ages could have hardly any traits in common. One of the most important aspects of educational progress in the last quarter century is that concerned with the integration of lower and higher school grades all along the line to conform with the natural development of the child.

This growing recognition of law governing mental development has caused the psychologist to turn his attention to the possibilities of prediction. Formerly the parents of a defective child were encouraged to hope that the child would outgrow his backwardness, and that, in general, a child's present abilities and behavior tendencies have little significance for later development.

We know now that in the vast majority of cases dull children become dull adults, average children average adults, superior children superior adults. As for the feeble-minded, they always remain feeble-minded. All the available scientific evidence points to the conclusion that the child of two years whose mental development is 40 or 50 per cent retarded will, in most cases, continue to be in the neighborhood of 40 or 50 per cent retarded. The child who has superior special abilities does not ordinarily lose them. Even the child's trends of interests are now regarded as having prognostic value. Educational and vocational guidance are thus gradually being placed upon a solid foundation. There is probably nothing in the child's mental life or his behavior that would be found altogether lacking in significance for later development if we knew all the facts. There is reason to believe that both delinquent and psychotic tendencies of adults have their prodromes of symptoms in childhood and that it may sometime be possible to identify in early life a majority of those who are headed for these dangers.

The relation between mental and physical growth has been extensively investigated. Correlations have been made between mental development on the one hand and growth in height, weight, lung capacity, motor ability, and physiological or anatomical maturation on the other. The correlations have usually been found positive but extremely low. There is only the slightest tendency for physical development to parallel mental development. Height, strength, dentition, skeletal ossification, and even pubertal changes bear little relation to mental growth. One may assume a close correlation between mental development and growth changes in the finer neural structures, but neurologists have not yet been able to tell us much about these correlations except in their grosser aspects. Glandular and other chemical influences upon mental growth and behavior patterns offer one of the most promising fields for future research. Cretinism, long known for its disastrous effects upon mental development, is probably only one of many such influences, though the most spectacular one in its obvious outcomes. Chemical correlates of personality types have been mentioned in an earlier section. Already the clinical psychologist and the physiological chemist are working side by side in the study of problem children.

Development is more dependent upon natural maturation than upon formal training. Gesell has shown that if one member of a pair of infant twins is for several weeks given frequent practice in a motor activity, such as climbing stairs, while the other member of the pair is given no practice, the latter, when later given the opportunity to practice, reaches within a few days the level of skill the other has taken weeks to acquire. Much the same seems to be true of mental acquisitions. The nine-year-old of average intelligence who has not been taught to read can be made ready for the third school grade within a few months. The twelve-year-old of average intelligence who has been deprived of opportunity to learn to read can be made ready for high school by the time he is fourteen. Researches in child psychology have shown that much time is wasted in schools by presenting subject matter prematurely. Many a child agonizes for months over long division who, a couple of years later, would be able to learn it in thirty minutes or an hour. The same is true of every type of curriculum material. Refinement of pedagogical method is often resorted to where timeliness of instruction would be far more economical and effective. Experiments have shown that it takes an extraordinary amount of pedagogical ingenuity to teach a strictly average five-year-old to read as well as first grade children are expected to, but that almost any method works fairly

[118]

well when the same child has reached the mental age of six and a half or seven years. It is not intended to convey the idea that nearly all school instruction is given prematurely. For the average child it is, in the main, fairly well timed; it is the exceptionally dull and the exceptionally bright who suffer. By the lock-step system the dull are promoted beyond their ability to master the subject matter, while the bright are under-promoted. The child of 75 I.Q. cannot master first grade work until he has reached the age of eight or nine years. The child of 125 I.Q. can do so at the age of five; yet by custom, and in many states by legal requirement, both are expected to enter upon first grade work at the age of six. In spite of unpsychological legal enactments, great progress has been made in recent years in adjusting the difficulty of school work to the child's maturation level, and further progress in this direction may be expected.

The question has often been raised whether there is any definite relationship between childhood brightness or dullness and the age when ability ceases to improve. It has been found that mentally defective children usually reach their maximum mental development a little earlier than do average children, idiots earliest of all, and next the imbeciles. However, above the level of defectives it is doubtful whether much correlation exists between grade of mentality and age of reaching maturity, although there is some evidence to suggest that the intellectually gifted tend to continue their mental growth somewhat later than average children do.

The Psychology of Learning. Learning has been one of the favorite fields of psychological investigation since the pioneer work by Ebbinghaus in the eighties of the last century. The greater part of this work has been devoted to investigations of memorizing and relatively little to learning to solve problems. The laws of use and disuse

[119]

were laid down by James in his "Principles of Psychology," and to these Thorndike and others have added the law of effect. By the law of use is meant that a mode of response is stamped in by repetition; by the law of disuse, that it tends to fade out when repetitions cease. The law of effect refers to the supposed stamping-in effect of the satisfaction or dissatisfaction which the response brings.

All of these laws have been called into question by recent investigations. It has been shown that the alleged influence of repetition does not always operate and that mental and psycho-motor functions sometimes continue to improve after practice ceases. As for the law of effect, it is difficult to understand how satisfaction which *follows* a successful response could operate retroactively to strengthen the bonds which have already acted. The law of trial and error, which alleges that learning to solve problems is typically accomplished by the gradual elimination of unsuccessful responses, at least in the case of animals and young children, has also been called into question on the basis of experimental data.

Though the laws of learning which once seemed to have been so definitely established are in dispute, learning nevertheless takes place. The mistake was probably in assuming that the laws were more absolute than they actually are, whereas a given law operates only at times and under certain conditions. As originally formulated and as usually understood, they make learning a more mechanical process than it in fact is, and postulate a neural mechanism far too simple. The investigations of Lashley and Spearman point to the conclusion that in learning the brain does not operate merely as a very complex set of bonds or connections, each serving a specific purpose, but that large areas of the cortex act to a greater or less extent in a unitary manner. The center of attention is shifting from the mechanical aspects of learning to the rôle played by insight. In the newer terminology, learning takes place by the formation of new Gestalten, or meaning complexes, not by the mere addition and subtraction of bonds. It is more a rational and less a mechanical process than was formerly supposed.

The preceding statements will doubtless seem very vague to the reader who is not acquainted with the recent literature of Gestalt psychology. However, the applications of the newer theories to education are fairly clear. The psychologist is coming to view the educative process less and less as mere habit formation, and to lay greater stress upon the part played by insight, attitudes, and interests. The more recent textbooks on educational psychology give less space to economic methods of memorizing and more to explanations of how meanings develop. The effect upon school work should be to reduce the emphasis upon routine drill and to give larger place to activities of the kind that stimulate the child to set problems for himself and to seek rational methods for their solution.

Because of limitations of space I shall omit entirely from this discussion consideration of certain issues which have figured so largely in texts in the psychology of learning: the form of the learning curve and of the curve of forgetting, the relative advantages of learning by parts or by wholes, and the economic distribution of practice periods. These issues have their interest, but they have to do chiefly with habit formation, which is the least important kind of learning.

A very controversial problem in connection with learning is the extent to which the effects of training are capable of transfer to other functions. This is the so-called theory of formal discipline. Until a half century or so ago it was almost universally believed that extensive transfer takes place, and that one of the main purposes of education should be to train such hypothetical faculties as perception,

[121]

reasoning, imagination, etc. This view was vigorously combated by Herbart on theoretical grounds, and later by James and Thorndike on the basis of experimental findings. The influence of these psychologists brought an extreme reaction to the older views and gave prevalence to the dogma that transfer is entirely too negligible to merit being taken into account in the choice of subject matter for the school curriculum. The weight of opinion of to-day favors an intermediate position. It is certain that the earlier educational theorists laid entirely too much stress upon mind training. It is hardly less certain that Thorndike and his followers have overestimated the specificity of training. The latter viewpoint favors selection of curriculum material almost entirely upon the basis of so-called "practical" values, to the neglect of values that have to do with the cultivation of intellectual attitudes.

Chapter V

PSYCHOLOGY IN INDUSTRY

by WALTER V. BINGHAM

INDUSTRIAL psychology touches life at many points. Most
of us spend a large part of our waking hours at work.
rom choice or from necessity, we are employed in one way
Fr another, in production, business, or personal service.
We naturally want our hours of labor to yield their fullest
return, not only of money wage, but of other values as well.
To increase the zest of accomplishment, the sense of the
worthwhileness of one's efforts, and the approbation of
one's fellows for work well done, are aspirations no less
universal than the desire for more money. These purposes
form some of the main objectives of industrial psychology,
a practical science of behavior dedicated to the increase of
both working efficiency and satisfactions.

What Industrial Psychology Has Aimed to Do. The aims of
industrial psychology may be viewed from both sides of
the shield. It has sought removal of sources of nervousness,
irritation, and discontent; elimination of needless fatigue;
banishment of a sense of inadequacy or futility; and release
from occupational fears, whether of wage cuts, unemployment, health hazards, loss of prestige, old-age dependency,
arbitrary supervision, or other sources of apprehension
connected with the work. And many of these objectives
may be stated positively, as efforts to find better methods
of occupational training, of supervision, of organization
of work, of hiring and placement, and of controlling the
working environment in the interest of greater comfort,
health, and earning power.

[123]

Outposts of progress in industrial psychology are scattered to-day throughout many sectors of this wide area. Our industrialized social order has inevitably faced questions such as: the capacity of mature men and women to learn new skills and to adjust to a fast-changing situation; the relation between ability to do work and interest in that work; the effects of monotony, boredom, and fatigue on the satisfactions and the earning power of industrial workers. Psychobiological as well as economic roots of industrial unrest have been unearthed. The disastrous consequences of anxieties due to feelings of insecurity on the part of wage earners, supervisors, and executives have been emphasized afresh in each recurring business depression. The techniques of personal relationship—between foreman and worker, between groups of employees and of managers—have called for thoughtful study and industrial experimentation, the better to achieve individual competence, as well as integration of purposes and resolution of conflicts, in the common interest. To attack such problems, educational psychology, social psychology, and psychopathology have been called upon, and—most of all—the psychology of individual differences.

To what degree are people better adjusted to their work to-day than heretofore, more genuinely contented, more productive, more resourceful? To what extent are they finding greater satisfaction and self-realization than would be the case if such problems as these had never engaged the attention of industrial psychologists? No one can say. The aggregate results of their efforts to date are admittedly much smaller than they ought to be, for technopsychology is very young, and the psychological problems of work adjustment, like other biological problems, are complicated with numerous variables not easy to control. In the following pages, however, instances will be described in which such problems have been met and mastered, illustrating in this way points of view, methods of investigation, and results of research, but leaving it to the reader to pass upon the practical social value of these investigations, or to project into the future the benefits which may ultimately accrue from a scientifically grounded understanding of man's behavior in relation to his daily work.

First, let us take a hasty glance over the vast field into which, here and there, industrial psychology has been pushing its outposts of discovery. We can then look more closely into the specific aims of recent research, comparing them with the aims of scientific management on the one hand and of mental hygiene on the other. We shall inquire particularly into the ways which have been used to measure attitudes, interests, and abilities, to validate employment tests and other aids in vocational selection and placement, to diagnose and correct occupational shortcomings, such as lack of skill and susceptibility to accident, and to learn the true nature of industrial fatigue; for these are some of the more strategic areas in which industrial psychology has been exploring.

A Glimpse of the Field as a Whole. Research in industrial psychology, as in other branches of scientific inquiry, consists essentially in finding out precise *relations between variables*. The variables which must be analyzed, described, measured, and related, are:

(*a*) Significant aspects of industrial behavior, and

(*b*) Factors conditioning this behavior.

How, for instance, is a worker's productivity best defined and measured? How does the scientist measure the satisfactions men experience in their work? What are the incentives, financial and non-financial, which lead workers to put forth their full efforts? And just what is the relationship between each of these different kinds of incentives, and the accomplishments and satisfactions which follow from their use?

The vast industrial experiment now going forward in Russia is challenging the hypothesis that monetary reward is an indispensable incentive. Was Lenin correct? Another decade may tell.

In America many industrial experiments are in progress, forced upon us by business conditions: experiments in resorting to part-time employment instead of layoff, and in maintaining wage rates instead of reducing hourly pay. Just what is the relation between reduced hours and the feelings and attitudes of employees whose working time is shortened? What are the mental and physical consequences of a feeling of security, or of insecurity? Most of these experiments, unfortunately, are being tried without any thought on the part of management that they furnish precious grist for scientific research, or that their outcome, scientifically measured, might yield conclusions invaluable to industry and to society. But here and there, for example in the mills of the Kimberly-Clark Corporation, serious attempts are being made to assemble scientific data on just such questions as these.

An apparently simpler problem in ascertaining the relationship between significant variables in the work situation is illustrated by an inquiry into the relation between intensity of illumination on the one hand, and output, accidents, and employee morale on the other, which has been made under the auspices of a committee of the National Research Council, and is soon to be published. This relation is not so clear cut as earlier experiments had seemed to indicate, since personal and social factors as well as physical aspects of the environment have had to be controlled before effects of lighting on industrial behavior could be unambiguously ascertained.

The variety and complexity of the factors which come within the scope of the industrial psychologist's inquiry are seen at a glance in the accompanying partial list of variables, in which managers and workers are chiefly interested.

PSYCHOLOGY IN INDUSTRY

THE FIELD OF INDUSTRIAL PSYCHOLOGY
Variables to Be Measured and Related

A. *Measurable Aspects of Significant Industrial Behavior*

Output

  Quantity of output
  Quality of output
    Proportion of spoiled work
    Excellence of product
    Variations in output

Earnings

Savings

Absenteeism, lateness

Labor turnover

Labor stability

  Length of service

Time required to learn

Rate of advancement

Health

  Medical examination data
  Records of illness

Safety

  Accident frequency and severity
  Lost time; minor mishaps

Suggestions—number and value of

Conflicts, individual and group
  Disagreements; emotional outbursts
  Acts of insubordination
  Restriction of output, sabotage
  Strikes, lock-outs

Fatigue (decreased capacity)
  Energy-cost (oxygen consumption)

Fatigue (feelings of weariness)

Interest in work
  Feelings of zest; absorption in task
  Feelings of boredom, distaste, unrest

Reveries
  Grouches, pessimistic ideas, worries, etc.
  Daydreams

Morale, labor attitudes

Satisfaction, contentment

as related to

B. *Factors Conditioning Behavior*

Fitness for work
  Individual abilities, characteristics, and desires, such as
    Age, education, and experience
    Interests; ambition
    Emotional stability
    Intelligence
    Strength and health
    Special aptitudes and talents

Social and economic status

Training

Supervision
  Methods and attitudes of supervisors

Organization of work
  Layout, routing, supply of materials, instruction cards
  Work methods, tools, machines, postures, variety, etc.

Hours, rest periods

Work surroundings: lighting, ventilation, noise, music, fellow workers, etc.

Financial incentives
  Salary, wages; fairness of rate
  Method of payment: day wage, piece rate, group bonus, etc.

Non-financial incentives
  Supervisory encouragement or drive
  Approval of fellow workers; honor roll, etc.
  Graphic record of production
  Group rivalry, etc.

Opportunity for advancement

Uncertainties concerning wage cuts, accidents, health hazards, old age, unemployment

Miscellaneous: personnel and management policies and methods; provision for participation in management, group insurance, unemployment compensation, etc.

Life outside of working hours
  Food, sleep, etc.
  Standard of living
  Use of leisure
  Home conditions

Each aspect of significant industrial behavior listed in column A, such as proportion of spoiled work, rate of advancement in salary or in responsibility, or degree of satisfaction in the job, has actually been measured or quite conceivably can be measured. The factors conditioning behavior, of which many are listed in column B, must also be measured, and their relationship ascertained to the behavior items in column A.

This is not the whole picture. Even when each of the humanly and industrially significant effects mentioned in the first column of the table is thoroughly understood in its relation to each of the causes listed in the second column, it still will be necessary to determine interrelationships. Quantity and quality of goods produced, for example, are determined in part, not only by the ability of the workers, their training, supervision, incentives, feelings of security, and other conditions listed in the second column; they are affected also by regularity of attendance, health, number and severity of accidents, and other variables appearing in column A. The task, then, is neither easy nor simple.

It does not follow, however, that the problem that faces the industrial psychologist is utterly baffling, and that the methods of science must, therefore, be put aside in favor of shrewd, unaided common sense, or that intuitive, impressionistic executive judgments based on conference and pooled "experience" are superior to precise records, measurement, and controlled experiment. True, the answer in its entirety is not going to be found in this generation, or the next. But already we know that, taking the problem bit by bit, exploring minutely the relations of one of its variables to a few of the others, the findings are often of immediate practical value. They are also steps in advance for the science of industrial psychology.

Unexpected By-products of Industrial Research. Current industrial investigations of the Western Electric Company in their great manufacturing plant at Hawthorne, a suburb of Chicago, are in point. The story is told in the *Personnel Journal* for February, 1930, by G. A. Pennock, M. L. Putnam, and Elton Mayo. These studies were initiated nearly four years ago, when a small group of women relay assemblers were separated from the other employees in this department, and a series of observations and experiments was begun for the purpose of accurately determining individual variations in output and the relationship of these ups and downs to conditions of work—particularly to such factors as method of payment, length of working day and working week, nutrition, sleep, length and distribution of rest periods, and the like.

The procedure required that conditions be maintained as nearly constant as possible for a period of weeks, followed by the introduction of a single change, such as provision for a light lunch at the time of the mid-morning rest pause. After a while another change was introduced, such as shortening or lengthening the working day. All this time, each worker's output was automatically recorded minute by minute. There was no pressure to speed up, no driving by the supervisor. But the workers were encouraged to tell how they felt, to comment on what they liked and disliked about the situation, and also to mention anything that happened outside of working hours which might be useful in accounting for their fluctuations in working efficiency.

The outcome has been astonishing. Workers' earnings and satisfactions improved far beyond expectation, and in some degree quite independently of the changes made in physical working conditions. While information of real value regarding optimal number, length, and distribution of rest periods, mid-morning lunches, and similar variables was secured, the management attaches far greater importance to what this experiment has revealed regarding the

[129]

characteristics of effective supervision. Indeed, a systematic effort is now being made, through a program of employee interviewing and a new type of supervisory training, to extend throughout the works to all the 40,000 employees some of the benefits that were first brought clearly to light in this modest experiment. Here, as in many scientific researches, the unexpected by-products have far exceeded in value the direct returns, important as those have been.

Gradually industry is realizing the great potentialities of such experiments as these. Management is carrying over into the realm of human behavior the same scientific ideals that it has long demanded of the chemist, the metallurgist, and the engineer. In so doing, it has taken steps toward the development of a well-rounded industrial psychology.

Relation of Industrial Psychology to Allied Fields. The aims of industrial psychology have a great deal in common with those of the mental hygiene movement on the one hand, and of scientific management on the other. When a psychiatrist approaches the personnel problems of a vast mercantile organization, he is not only plunged at once into those familiar problems of emotional maladjustment with which the physician of the mind has largely been preoccupied; he also faces many practical details of supervisory training and executive organization, not to mention the whole range of practices and techniques designed to select employees for particular duties and to adjust them to occupations in which their natural abilities and predilections find the greatest opportunity.

The expert in scientific management also, faced with the necessity of securing maximum output at minimum cost in order that the enterprise may prosper and continue to furnish steady employment, realizes that morale is essential, that excessive labor turnover is inordinately expensive, that industrial accidents due to worry and other preoccupations are an avoidable waste, and that the highest productivity can be obtained only with a vocationally well-adjusted personnel.

So it is that mental hygiene, industrial psychology, and scientific management have a great deal in common. What, then, are the essential differences? They lie partly in the relative prominence given to particular objectives, and partly in specific techniques and methods of approach, resulting in three somewhat overlapping and yet not identical bodies of knowledge, principles, and practices. Each is valuable.

The specialist in mental hygiene, for example, trained in medical school and hospital, experienced in healing the mentally ill and in preventing nervous breakdowns, has keen eyes for conditions predisposing to anxieties and bad emotional habits. He is on the alert to diagnose and correct situations that tend to mental illness. His aim is health. Increased productivity of workers whose personalities are already well balanced is distinctly a secondary consideration. It is the problem case that first engages his attention. His typical method is that of the intimate personal interview.

The scientific-management specialist usually comes to his task with the background and training of the engineering school. He knows mechanics, economics, statistics. His eye rests on the machinery and the layout, the sales index, the production chart, and the balance sheet. He emphasizes output. His familiar tools are the slide rule and the stopwatch. Far from ignoring the human factor, he sees it as one of several important terms in his equation. His knowledge of human nature, frequently sound and shrewd, has been gained from practical experience in dealing with executives and workers in the plant. When technical or obscure questions of behavior arise, he supplements his common-sense psychology by calling upon the industrial psychologist or psychiatrist, or by instituting controlled

[131]

experiments within the industrial situation and measuring the results. Indeed it is the management engineer who must be depended upon to see that the entire enterprise is so organized and administered that the staff services of industrial psychologists can function effectively.

The industrial psychologist puts the individual first, output or profits second. His training in the university laboratory and in office, store, or factory, where all sorts of workers and supervisors are employed, has impressed upon him the wide range of differences among people, in their capacities, tastes, and requirements. He knows how great is their susceptibility to training, even though they may be well on in years—provided this training is individualized and adapted to their separate needs. He has studied the springs of action and the laws governing acquisition of skill, modification of habit patterns, control of motives, and improvement of social adjustments. He, too, uses the interview, but has a predilection for checking its findings against other data and supplementing them with objective measures of performance. Indeed he is incorrigible in his insistence upon full personnel records and concrete, measurable facts. His favorite instruments are the reaction key, the kymograph, the test blank, the correlation chart. But his primary goal is not, like that of his academic colleagues, the advancement of understanding of general principles; it is effective adjustment of individual workers within their several work situations. He aims at steady increase of their earning powers up to the limits of their capacity, and at the contentment and satisfactions that can come only to those who are happily, because fittingly, employed.

The Physical and the Social Environment. Psychologists have taken their techniques and point of view into mines, factories, railways, advertising agencies, farms, printing establishments, restaurants, hotels, and aviation fields. They have studied workers and managers in textile mills, machine shops, telephone exchanges, banks, museums, libraries, laundries, and power plants. In government departments and public utilities, as well as in private industries and stores, they have analyzed work processes and the conditions that affect individual variations in performance. They have studied the abilities and aptitudes as well as the duties and difficulties of the men and women there employed. Improvements have been made so that the work could be done better, with less expenditure of energy. Appliances, tools, and benches have been adapted to the requirements of the human organism. Standards of ventilation and lighting have been modified, to remove discomfort and strain. In a wide variety of situations, the mental effects of physical working conditions have been scrutinized in the interest of greater comfort, convenience, ease, and accuracy of work.

Not only the material surroundings have received attention; the personal environment also has been studied—the psychology of supervision and of relations between fellow workers. Nothing affects a man's mental attitude more intimately than his contacts with his immediate superior. So techniques of training and of personal leadership have been investigated, and sound principles embodied in courses for supervisors, foremen, and managers, with the result that the general level of supervisory practice in industry is being steadily raised. Reasons for workers' restriction of output have been investigated. Obscure and unsuspected causes of slackness and indifference have been uncovered. The values of group incentives to teamwork have been demonstrated in particular instances. Attention has been drawn again and again to the possibility that familiar but often overlooked assets can be capitalized when workers are given a real opportunity to contribute their own ideas, share responsibility for those aspects of the enterprise closest to them, and feel a pride in its success. To be competent in one's job is indispensable; but to see this job in its total setting and to receive from one's fellows a recognition of its worthwhileness are also essential, if the will to work is to be fully released.

Such basic hypotheses about human nature in relation to both the physical and the social aspects of work, it has been the duty of industrial psychology to investigate and to apply.

Measuring Mental Attitudes, Interests, and Abilities. One way of bringing different components of the total work situation into correct perspective has been for the industrial psychologist to measure employee attitudes toward the firm's personnel policies and practices. More than seventy years have passed since Fechner laid the corner stone of experimental psychology by demonstrating the possibility of mental measurement and formulating the fundamentals of method; but only within the past decade have these methods been adapted and applied to the practical task of measuring such industrially important quantities as group morale, good will, and employee preference for various features of management practice. These techniques constitute a real contribution of psychology to industry.

A similar accomplishment of industrial psychology has been in the direction of measuring a man's interests and relating these to the requirements of various kinds of work. Here the psychologist must deal with the individual rather than the group; and since the single measure has a lower reliability, he uses it with full awareness of its limitations. The same may be said of measurements of other vocational aptitudes and proficiencies. The invention of standard tests of skill, trade knowledge, manual dexterity, mechanical ability, mental alertness, and capacity to learn, has reduced to some degree the hazards of predicting an applicant's probable success; but in only a relatively small fraction of the occupations has the validity of such tests

[134]

as yet been demonstrated, and even here such data must always be appraised in relation to all the relevant facts obtainable about the person's previous experience, social and economic status, success in school, health, temperament, emotional balance, and the like. Employment psychology, let me hasten to add, has provided a scientific procedure for determining what relative weight should be given to each of the several items considered, in hiring for those occupations in which the number employed is sufficiently large.

A man's success in an occupation is obviously conditioned by many factors, both internal and external. His determination to succeed, and his degree of interest in the particular work to be done, may be as crucial as is his ability to perform the tasks required. Can such aspects of human nature actually be pinned down and measured? An uninformed vocational interest is notoriously volatile. Interests shift with knowledge and experience. Deep-lying antipathies can sometimes be overcome. And yet it is almost axiomatic that satisfactory adjustments to one's work are enduring only when that work is of a kind that matches natural tastes as well as abilities. So industrial psychology has eagerly watched the development of scientific means for ascertaining fundamental preferences or bents. The road has been long and the end is not yet. Beginning with Miner's check list of occupational preferences for use in the vocational-counseling interview, the first milepost was a statistical comparison of likes and dislikes among groups of salesmen and engineers in Yoakum's seminar at Carnegie Institute of Technology ten years ago. Moore's research for the Westinghouse Electric Company, on the differentiation of graduate engineers into those who would eventually become successful engineering salesmen rather than designers or supervisors of production, is classic. Freyd developed a still better instrument and used it in his

[135]

studies of personality; and many others, notably Strong, have further refined and extended these methods of attitude measurement as aids in ascertaining occupational and professional interests. Such techniques have a place and will probably not be entirely superseded even though the psychologist at some future time succeeds in inventing still more direct and objective means of measurement, based on determination of what the individual actually does when confronted by a choice of opportunities.

Industry's Use of Psychological Tests. The successes of the military psychologists in 1917–1918, and during the days of the occupational-rehabilitation movement following the World War, served to spread throughout industry an acquaintance, even though superficial, with the aims and possibilities of employment psychology, giving focus to the general interest which Münsterberg had stimulated some ten years before. Many thousands of intelligence-test blanks were purchased for business use. Firms like the Eastman Kodak Company installed, as part of their selection procedures, peg-board performance tests of manual dexterity and other objective measures of ability, to help in discovering talent for inspection work as well as for assembly operations and jobs requiring mechanical ingenuity. The extent to which some firms have carried the use of tests as an aid in selection is evidenced by the fact that during 1930 the employment department at Macy's gave psychological examinations to more than 14,000 of the applicants for positions in that store.

The value of a rigorously scientific procedure in developing and adapting such tests to the demands of a specific situation is well illustrated by an investigation recently made by Viteles in the Philadelphia Electric Company. Here operating mistakes of electrical substation operators have been reduced 43 per cent through the use of psychological tests in the reassignment of these men. The importance of such a reduction is evidenced when it is recalled that a mistake at the great switchboard of a substation may deprive a hospital of electricity while a surgical operation is in progress, or result in interference with a manufacturing process entailing a loss of many thousands of dollars. Other examples of the same kind could be cited from industrial research both here and abroad.

The federal postal service has been much improved since 1923, through simplification of the work of mail sorters after psychological analysis of their job, and through replacement of the traditional type of civil service examination by a more practical, convenient, objective type of examination, by means of which the U. S. Civil Service Commission has annually tested some 60,000 applicants for employment as distributors, carriers, and railway mail clerks, more easily, accurately, and fairly than before, with much less expense to the government, and with marked improvement in the average ability of the men appointed. Principles and techniques which O'Rourke, Thurstone, and their associates developed in connection with such psychological research on employment procedures have since been applied not only to other civil service examinations, but to similar problems in business and in education also.

In the transportation field, Viteles and Mrs. Shellow, for the Milwaukee Electric Railway Company, developed and validated tests for selecting applicants to be trained as street-car motormen—men who can keep their minds on the job and do the right thing in spite of distractions and sudden emergencies. The work of Snow for the Yellow Cab Company, Segard for the Third Avenue Railway of New York, and Wechsler, Moss, and others for various transportation firms, paralleled in this country elaborate developments in the technopsychological laboratories of

the Paris tramways and the German railways. These
methods have proved valuable in reducing accidents chiefly
in those companies where the major executives have seen
that such selection procedures are properly integrated in a
well-considered program of training and individual supervision, as has been done among the delivery drivers of
R. H. Macy and Company.

The Western Electric Company, the Atlantic Refining
Company, and many other firms have extended into the
supervisory and executive levels their investigations of the
usefulness of psychological tests. Considerable effort was
devoted to the search for dependable aids in predicting
ability to sell, both in retail stores and on the road. But
the widest use of tests has been in the selection of typists,
stenographers, file clerks, comptometer operators, and
other office workers, following the pioneer work of
Thorndike for the Metropolitan Life, and of Scott for
Cheney Brothers.

The expectations of the uninformed that recourse to psychological tests could somehow relieve the employment
manager of the necessity of using also the more familiar
ways of sifting and placing applicants, were early dispelled.
It was also recognized that vocational adjustment is a
continuing process. It begins in the schools. Initial placement and replacement are incidents along the road of self-
discovery and advancement. The industrial psychologist's
interest, then, reaches back into the period of early vocational guidance, and continues throughout the worker's
occupational career. If, historically, this interest seemed
to find a locus first in the employment office, it almost
immediately reached out into the plant. The processes of
training on the job, bristling with problems essentially
psychological, early engaged the attention of pioneers like
Link, who had first entered industry to improve the
procedures of hiring. The mere necessity of knowing the

nature of the various jobs, and of getting dependable measures of the later occupational success of people hired, was enough to draw the psychologist out of his laboratory into the works. Miss Pond, of the Scovill Company, and Frazier, of Dennison's, for example, are typical of many who have begun by investigating employment tests and found themselves plunged almost at once into a consideration of supervisory relationships and problems of organization. Johnson O'Connor, of the General Electric Company, who for eight years has consistently held himself to research on employment tests, previously had had a background of industrial engineering and plant experience. The industrial psychologist cannot arbitrarily isolate his problems and concentrate his efforts on a single phase of the task of occupational adjustment without reference to the total setting.

Intelligence and Length of Service. The methods and results of employment psychology may be illustrated in somewhat greater detail by a few typical studies, the purpose of which has been to find the relationship between length of service on the job and intelligence test score, amount of schooling, and other variables obtainable at time of hiring.

Just what, for instance, is the relationship between length of service in routine clerical work and the intelligence of girls employed for such work? Can it be said that the brighter they are, the more stable they will be? Or does the opposite generalization hold, that the duller they are, the longer they stick to this kind of routine work? As a matter of fact, neither of these hypotheses is correct. Over and over again it has been found that the relationship between the two variables mentioned is not rectilinear. Scatter diagrams of intelligence test performance plotted against length of service show that routine clerical workers who are at once competent and contented, are more often

[139]

neither very dull nor very bright. Above an upper critical point on the intelligence scale, as well as below a lower critical score, the proportion of girls who disappear from the payroll is greater than it is within the middle zone of mental alertness. Many other variables, such as wage rate, also are associated with permanency of employment, so that this characteristic relationship between brightness and stability on a relatively routine job is far from close, but it is striking enough to be significant. Since this fact was first definitely established by Yoakum and Bills some ten years ago, it has been kept in mind by all who use measures of mental alertness as one of the aids to selection of office workers. The hypothesis was not new in 1920, but its verification was possible only after reliable ways of measuring mental alertness had been developed.

A clean-cut demonstration of an upper critical score for success in a selling occupation was made at Carnegie Institute of Technology in 1919. Three years previously a group of twenty-seven companies maintaining national sales organizations established the Bureau of Salesmanship Research, later the Bureau of Personnel Research. This bureau was to pool the experience of the cooperating members, to evaluate their current procedures, and to experiment with new ways of selecting and developing salesmen. The first year's work, under Walter Dill Scott, resulted in a manual of "Aids in Selecting Salesmen," containing an improved personal history record or application form, a model letter of reference to former employers, a guide to interviewing which helped the interviewer to focus his attention on essential traits and to record his judgments quantitatively, and a set of five psychological tests with directions for administering them. Among these tests was a group intelligence examination, a forerunner of Army Alpha. It was given to various groups of salesmen and sales applicants, and their scores were then compared with their actual success as measured by amount of sales. Among the men so examined were forty salesmen for a food products company. To the dismay of the investigators, when the test scores were compared with the men's sales production records, the correlation was almost zero. This seemed to be a severe indictment of the test as a measure of intelligence. Then came the war, and with it a vast experience in personnel classification and intelligence examining. The psychological tests proved their worth as indicators of mental alertness. So, when Major Yoakum, with his background of Army experience, assumed direction of the Bureau of Personnel Research in 1919, he knew that the intelligence test methods were valid, and he sought a fresh explanation of the riddle in the findings of 1916. Using the same data, he computed the correlation between test performance and length of experience with the company. The correlation was not zero. It was negative: −0.40. In other words, the brighter the salesman, the sooner he tended to leave the employ of that concern. Yoakum repeated the experiment with seventy-six salesmen of the same company, using the best available adult intelligence examination. The correlation between test scores and length of experience was −0.46. A job analysis showed that the work required of these men was largely routine order-taking. The pay was meager. Chances of promotion were slight. Only plodders were content to remain long enough to get necessary experience and build up a creditable sales record. Examining the intelligence scores again, it was apparent that there was an upper limit as well as a lower limit within which the chances were large that an applicant for a position with this concern would make good. Below this zone he lacked the necessary mental ability. Above it the probabilities were that he would not be content to remain long enough to learn his work thoroughly. The earlier form of psychological test had, after all, been a reliable measure of mental alertness. The need had been for precise determination of the relationship between test scores and occupational success.

The preferred range varies for salesmen of different kinds of products and also with the territories within which they operate. In many occupations it has been shown that there is no upper limit to the optimal intelligence score; but studies of policemen and of machine operatives, as well as of salesmen and of clerical workers where the task is essentially routine, have shown how necessary it is to keep an eye on the upper as well as the lower critical score, in order to avoid anomalous or ambiguous inferences.

Statistical Methods in Employment Psychology. In developing aids for selection and placement of workers, the statistical method of *group differences* has proved useful, particularly in evaluating items of personal history information which throw light on the applicant's character, temperament, and emotional adaptability to the work for which he is being considered. The method is simple enough in application.[1] The groups compared are the successful and unsuccessful workers employed at the occupation in question; or those successfully engaged in the work, compared with a sample of the population at large. The proportion of each group answering an item on the application blank in a certain way, or making a certain critical score on a test, is ascertained. The difference in proportions is then computed, and also the standard error of this difference. If the difference in proportions is more than twice the standard error of the difference, it is considered to be significant for this purpose. Then the individual applicant's performance with reference to the item in question is given a weight corresponding to the size of this ratio, in the total score. Or, if the number of questionnaire items or test scores to be

---

[1] BINGHAM and FREYD, "Procedures in Employment Psychology," Chaps. XIII and XV, McGraw-Hill Book Company, Inc., New York, 1926.

treated in this way is large, not much reliability is lost if each item which proves to be significant is given a weight of +1 or of −1, as the case may be, in computing the total score.

Such a method serves to add to the value of items of information obtained on the application form or in personal interview, regarding age, schooling, previous experience, marital status, and many similar considerations which may or may not be significant indicators of probable success. It lifts the evaluation of such items out of the area of guess-work or subjective impression.[1] The method is equally applicable to items obtained in the physical examination, such as height, weight, eyesight, strength, and blood pressure.

Preconceptions as to the importance of such facts about an applicant have sometimes been modified or even reversed when sufficient data have been gathered and the computations made. As an illustration may be cited a study of a group of young women operators made by the writer in one of the plants near New York City. They were engaged in tending machines which wind paper insulation about strands of copper wire for making telephone cables. These machines make a deafening noise, yet the girls work here month after month and do not seem to mind it. Their minimum wage is thirty-eight cents an hour, but the most skillful operatives among them are able to increase this rate to as much as sixty or sixty-five cents. Even during the period of learning the work they receive a better wage than the average high-school graduate who goes into an office as a typist, and after two to eight months the more competent ones earn more than a college graduate usually does during her first year in an office. The work is not

[1] MANSON, GRACE E., What Can the Application Blank Tell? Evaluation of Items in Personal History Records of Four Thousand Life Insurance Salesmen, *Journal of Personnel Research*, Vol. 4, pp. 73–99, 1925.

wholly unlike that of tending spindles in a cotton mill. It calls for some skill, strength, and mental alertness; but it seems to require in addition a certain stolidity and poise which enables the operator to keep steadily at work when the paper breaks on two or three heads at about the same time. If she is high-strung, she becomes nervous or excited under these circumstances, so that matters go from bad to worse; whereas if she is not tense or irritated, she gets out of trouble more quickly. Her output is less likely to suffer. She makes better wages, and is apt to be more contented and to continue at the work.

The problem of vocational selection for such a job pointed toward the more precise utilization of information already being gathered on the application form, in personal interview and in the medical examination. A check-up of items was made in order to determine which were really significant in sorting the successful from the unsuccessful operators. For purposes of this study an operator was classed as successful if she proved to be both able and willing to stay on the job for six months. An unsuccessful operator was defined as one who disappeared from the payroll within that period. The main group studied consisted of all operators hired during a period of sixteen months, totaling 246 young women. Of these, 43 per cent became successful operators in the sense indicated. Data regarding them were drawn from the tabulation cards of the employment department, from the application blanks, and from the files of the medical examiner. For each item for which information was recorded, a comparison was made between the successful and unsuccessful operators. Wherever differences appeared, the percentage of each group falling within the preferred range was computed and the significance of differences in percentages determined.

On most of the items there was no discoverable significant difference between the successful and the unsuccessful operators. For instance, we could find no preference for tall girls or short girls, or girls of medium height. There was about the same proportion of any particular height in the unsuccessful as in the successful group.

In age, on the other hand, there was a preferred range. Girls nineteen years of age or younger at date of hiring were a little more likely to be successful than those twenty years old or older. The difference was not great. About 50 per cent of the younger ones succeeded and about 34 per cent of the older ones.

As to schooling, the data did not give clear-cut results. Few candidates with high-school training had been hired. Many had not even completed grammar school; indeed, the preference was slightly in favor of those who had not finished the grammar grades.

As to average length of time on previous jobs, we tried various ranges and finally found this preferred range: if the average was less than one year, it was an unfavorable indication; if it was one year or more, the indication was favorable.

So we went through all the items available. Most of the findings simply confirmed impressions that the employment interviewers already had in mind, although they may not have been giving as much weight to some of the items as they did after their actual statistical significance had been determined.

As to nationality, however, some interesting results appeared which had been anticipated only in part. Using the data on the application blank, we classified the employees in five groups according to racial stock. In the first group were the Austrians, Germans, and Scandinavians. Of these, 63 per cent were successful. Then came the Czechoslovak and Polish, the largest group, of whom 50 per cent were successful on the job. Of the Italians, 43 per cent were successful. Then the Irish, English, and

Scotch, that is, the foreign English-speaking stocks, of whom 26 per cent were successful. Of the Americans (defining Americans for this purpose as those whose fathers were born in this country) only 19 per cent were successful in this occupation.

These differences are striking. Instantly there occur many possible explanations, social and educational, as to why American girls in that community prefer to take lower wages working as clerks, than to earn better wages at such factory work as this. Whatever the reasons, here is an instance in which the statistical method applied to familiar data from the application interview, revealed information significant for predicting likelihood of success, thus supplementing the ordinary procedures of hiring and placement. Such aids can be employed only when there is a sufficiently large number of employees in a single occupation, and when reliable, clean-cut criteria of success in that occupation are available. Probability tables based on data of this sort have been compiled from their own extensive experience by several firms maintaining large sales organizations and by groups of life insurance companies whose expenditures in recruiting and training successful salesmen of life insurance constitute a measurable item of expense to the purchaser.

It is not only data of the kind described which lend themselves to useful purposes when gathered carefully and subjected to quantitative evaluation. The interest questionnaire or occupational preference blank also is used in this way.

Somewhat similar approaches to quantitative determination of professional aptitudes are those which resort to the free association experiment and similar sources for data regarding temperament, tastes, and emotional attitudes. In this field of inquiry, Kenagy and Yoakum, Freyd, Wells, Laird, and O'Connor may be mentioned as among those who have not neglected the application of

[146]

these techniques to industrial situations. Behavior data, which may be gathered quite incidentally to the administering of tests of intelligence or skill, are another source of information regarding significant differences of temperament and attitude. Finally, the personal interview, skillfully administered, may be made to yield suitable grist for the statistical mill.

The employment psychologist or the vocational counselor has done but part of his task when he has developed and applied suitable measures of ability to do or to learn to do the particular work in question. He needs also reliable techniques to aid in predicting probable contentment and satisfaction in that work. We have seen that predictions may in many instances be based on statistical evaluation of data already at hand, but too often neglected, or used only in an intuitive or common-sense way. Such data are found on application forms, physical examination records, and preference questionnaires. They may be gleaned as by-products of the personal interview or of intelligence and trade test performance. They should be used with discretion, after their reliability and their validity for the particular purpose have been statistically determined.

Psychological Aspects of Safety. Ours is a dangerous age. The tempo of modern living has been accelerated far beyond anything imagined by our fathers. The pace of traffic, by land, sea, and air, matches the increasing speed of factory machines. The resulting hazards have challenged the ingenuity of engineers, who have invented clever protective devices, guards, and automatic controls. But in spite of the perfection of these mechanical aids to safety, serious accidents have continued to occur in great numbers; for failure of the human factor causes far more mishaps and disasters than failure of a mechanism. And so industrial psychology has sought, with gratifying success, to make its contribution to safety.

[147]

"The best safety device," my friend Bill Pfouts used to say, "is located above the neck." What has industrial psychology done to improve the efficient operation of this device?

The story would be long, if all were told. Distinguished psychologists in Germany, Russia, France, Denmark, Italy, and other lands have delved into the mysteries of motive and habit, attention and distraction, visual acuity, reaction time, susceptibility to fatigue, self-control, and other aspects of human. nature, in search of obscure causes of proneness to accidents and ways of eradicating them. Our colleagues of the British Industrial Health Research Board have studied large numbers of apprentices and employees in shipyards and factories, to determine the relation of accidents to monotony of work, to proficiency, and to various differences of ability and personality. Their statistical investigations early proved that industrial accidents do not just happen; they are not distributed among the workers according to the laws of chance. In this country also, in studying records of accidents among factory employees, street-car motormen, automobilists and bus drivers, more than half the accidents have been found to occur to a relatively small proportion of the men. These accident-prone workers are, for the most part, as eager to avoid accidents as any of the others. They have had the same training and supervision. They see the same posters and take part in the same safety drives. So the problem of the industrial psychologist is clearly that of developing the most effective ways of studying these accident-prone individuals and helping them to overcome their particular proclivities.

Accidents have been largely reduced where this psychological approach to the problem has been added to the more familiar forms of effective safety effort—on the street railways and bus services of the Boston Elevated Railway, for example, where the writer personally has had opportunity during the past three years to cooperate with the management, study the problem in detail, and observe the benefits to employees and public that have followed the use of the procedures recommended. Here, continuing studies have been carried forward, and practical procedures developed and installed. Collision accidents have been reduced more than 35 per cent. Men and management are proud of their fine accomplishment; and public good will toward the road has been increased among the car riders and pedestrians of the metropolitan district which it serves. The financial saving effected through reduction in deaths, personal injuries, and property damages has exceeded $300,000. Industrial psychology has its economic as well as its humanitarian values.

An Investigation in the Interest of Safe Transportation. How have these results been brought about? In what ways have the methods and results of psychological investigation helped in solving this practical problem of conserving life and limb?

The task has not been a simple one. The motormen and bus operators, about 2,500 in number, of necessity do most of their work without immediate supervision. The hazards of the city traffic in which they operate are varied and have been continuously increasing. Collisions, both trivial and serious, between street cars or busses, and automobiles, other vehicles, or persons, number several thousand each year, and cost the company upwards of $1,000,000 annually. This cost has tended to increase since the war, owing not so much to increase in number of accidents as to the greater cost of settlements. Juries have been giving larger and larger verdicts for damages. Before our cooperation with the company began, they had been able to keep down the number of accidents by using the commonly approved methods of prevention, such as mass education, safety posters, and group rivalry, as well as adoption of mechanical safety devices; but in spite of these efforts the costs were heavy, amounting to more than 3 per cent of the gross receipts. In most other large cities, it should be noted, this percentage is even higher. No wonder the management was concerned to decrease the frequency of accidents.

The question was raised by the general manager and the claims attorney as to whether accidents might not be reduced by a more careful selection of men to be trained as motormen. It was remembered that in 1908 Münsterberg had broached the subject of individual differences among motormen and, after working for some time in his laboratory to devise suitable tests, had examined six of their experienced men—three with a high accident rate, three with a low. His tests did not differentiate perfectly between the two groups, and he retired to revise the methods. His writings on the subject later stimulated much research. In Paris, Berlin, Prague, Milwaukee, and elsewhere, such tests as he proposed were developed into valuable aids in the selection of motormen. So the management of this particular railway called on the Personnel Research Federation for counsel, and the present investigation was begun.

Our approach to the problem, however, was made along other lines. Instead of focusing on initial selection we decided to study the experienced men who had had more than their share of accidents, to see whether ways might not be found to help them to become safe operators. It is important to keep in mind that the results achieved as a consequence of adopting this approach have been attained by methods other than better selection, or by discharging or transferring accident-prone men.

The foundations for our work were laid in a comprehensive job analysis, accident-location studies, and a series of statistical investigations. One of the first of these statistical inquiries showed a positive relation between accident proneness and high blood pressure. Of course no man known to have excessively high blood pressure is permitted to operate a street car; the danger from a stroke is too great. The men we studied had a higher than normal, yet not dangerously high, blood pressure; yet during 1926 they had had two and one-sixth times as many collisions as other men of similar age and experience. Significant associations were also found between freedom from accidents and such variables as age, experience, and operating efficiency as measured by automatic recorders of "coasting time," the time when neither brakes nor power is applied.

Studying the Motorman and His Job. To make an analysis of the motorman's occupation, I myself operated a street car. Another member of the staff operated one until he had an accident. He also, with stop watch in hand and a stenographer by his side, observed in detail the operation of many motormen in starting the car, "notching up" the power, approaching intersections, rounding curves, crossing switches, responding to signs of danger in complicated traffic, applying the air brakes, opening and closing doors, calling station stops, answering questions, making change, helping cripples or fat old gentlemen on and off, changing ends at the terminus of the run, setting the proper signs, inspecting equipment, replacing a burnt fuse, and meeting the many emergencies that arise from time to time while operating in congested traffic. Still another staff member went through the regular course of training, making a full record of duties, activities, and difficulties.

Among the many points brought out by these studies it is sufficient here to cite a few. A motorman needs to be capable of distributing his attention broadly, so as to grasp the whole changing situation. He must know and recognize promptly minute signs of possible danger, such as a little smoke coming from the rear of an automobile parked ahead of him beside the tracks. He must not be easily susceptible to distraction or to panic. There are certain skills and correct job habits which need to be thoroughly automatized. In many respects the requirements are similar to those of the crane operator, the aviator, the locomotive engineer, the chauffeur, or any one who has to control a moving machine, manipulating handles by means of movements which are relatively simple but which must be made at the right time, quickly and correctly, in the midst of a changing situation.

We fitted up a modest psychological laboratory near the offices of the medical examiner and the supervisor of employment and training. When men came to take a physical examination, some psychological tests were also given, designed to measure speed and accuracy of response, consistency of serial action, flexibility, oscillation, perseveration, resistance to distraction, range of peripheral vision, and similar attributes. Some of the tasks consisted of pressing the right buttons as various electric lights flashed momentarily. Other tests made use of the familiar street-car controls—the electric power handle, air-brake lever, gong, etc.—all connected with suitable recording apparatus.

These psychological tests were used for research and as clinical aids to supplement, in certain instances, the analysis of a man's difficulties made by studying the detailed descriptions of his accidents, interviewing him, and observing him at work. A number of supervisors were specially trained to ride with the motormen and note the points on which they individually needed further training or supervision. A man's accident record was first examined to see whether there was a particular type of accident to which he was prone. One man had had five collisions in as many months. The supervisor who observed him in action reported that he was competent, skillful, and careful.

A comparison of the five collisions showed that they were all alike in two respects: they were all of one relatively infrequent type, namely, front-end collisions; and all five of them happened on Sunday afternoons. Another search of the records showed that all the various cars operated by this motorman through the week were medium-weight cars, but his schedule on Sunday afternoons called for operation of a large, heavy car. Other motormen with similar schedules had no accidents, but this man lacked the necessary flexibility of habit. In other words, he was a perseverator. This peculiarity of mental make-up is measurable in the laboratory. After his schedule was changed so that all his cars were medium weight, he stopped having front-end collisions.

Consideration of the details of a man's accident record, observation of his actual operation, and personal interview with him all help in diagnosing his difficulties. For instance, a man of sixty, with an excellent record over many years, began to have a good many collision accidents. Reference to the records showed that eight out of eleven of them happened just as he was starting his car. Was he simply getting careless? Not at all. Observation and inquiry showed that he was so conscientiously careful about one part of his work that it was interfering with another part. He had become slightly deaf, and had formed the habit of listening for the starting signal from the conductor so intently that he neglected to look about in front of his car before throwing on the power. By a little retraining he learned to look quickly from side to side instead of starting his car without doing so. Since that time he has had no accident when starting.

Some of the cases of accident proneness turned out to be medical, some were traceable to wrong mental attitude, while others needed specific information or training. No two cases were exactly alike, so it was necessary to study them individually and to deal with each man differently in order to help him to overcome his suceptibility to accidents.

Although each case of accident proneness is unique and complex, and the essence of our procedure in combating accidents is *individual* attention, we have made certain group comparisons from time to time, to determine the relationship of various isolated traits to the practically important trait of accident susceptibility. One special report of a study by C. S. Slocombe and O. M. Hall covers eighty-six motormen, half of whom have had more than the average number of accidents. Forty-three of them, then, we may call high-accident men; forty-three, low. They were selected and paired for age, length of service, and operating conditions. These two groups, differing in ability to avoid accidents, have been compared with reference to twenty-four other variables. The results are summarized in the accompanying table.

Reference to the table shows that among these eighty-six motormen, seven had hernia, and all of these are in the high-accident group. Sixteen had a somewhat abnormal blood pressure, and fourteen of these are found in the high-accident class. Of eight who did relatively poorly in a serial-action test, all but one are high-accident men. Thirteen had on their records instances of insubordination; of these, two were in the low-accident group; and so on.

Here is evidence that accident proneness is associated with personality defect or uncooperative attitude, as measured by records of insubordination, operating delinquencies, and failure to report for duty. Thirty-nine per cent of the high-accident men, but only 5 per cent of the low-accident men, were defective, as measured by this record of uncooperative behavior. Forty per cent of the high men and 12 per cent of the low showed lack of aptitude, as measured by test S. Forty-nine per cent of the

[154]

PSYCHOLOGY IN INDUSTRY

ASSOCIATION BETWEEN ACCIDENT PRONENESS AND VARIOUS CONDITIONS
Among Forty-three Motormen with High Accident Records and Forty-three with
Low Accident Records, Paired for Length of Service, Age, and Operating
Conditions

Condition present	High	Low	Coefficient of association, Q
Hernia.	7	0	1.00
Abnormal blood pressure.	14	2	0.82
Any health defect.	21	4	0.81
Psychological trait S (score 3 or more).	7	1	0.81
Insubordination (one or more reports, 1926–1928).	11	2	0.75
Psychological trait S (score 2 or more).	17	5	0.66
Operating delinquencies.	18	7	0.57
Misses.	11	4	0.54
Set-backs refused.	10	4	0.49
Absences.	23	14	0.41
Overweight	12	6	0.41
Psychological trait G (score 10 or more).	12	7	0.33
Delinquency record (signs, etc.).	7	4	0.31
Psychological trait G (score 5 or more).	20	14	0.28
Complaint by patrons.	13	11	0.12
Dullness (observed during testing).	9	7	0.12
Record of shorts and overs.	13	12	0.04
Overtime (earnings over $2,300).	11	12	0.06
Breaks in employment.	7	8	0.08
Less Reliable Items:			
Drinking.	3	1	0.52
Economic status (chiefly trusteed wages).	5	3	0.27
Defective vision (not serious).	5	3	0.27
Difficult home conditions.	3	3	0.00
Medical not specified above.	3	3	0.00

Explanation of Table:
  Psychological Trait S means relatively poor performance in a serial-action test.
  Insubordination means a recorded instance of insubordination within the past
    three years.
  Delinquencies means a record of operating ahead of time, not slowing up at
    street intersections, etc.
  Misses means failure to report.
  Set-backs refused means refusal of operator to follow regulations when he has
    been long delayed in arrival at the end of his run.
  Psychological trait G means a group factor: relatively poor performance in the
    whole battery of tests.
  Shorts and overs means turning in more or less money than is shown by the
    register.
  Overtime means a large amount of overtime work as determined by reference to
    earnings. (Motormen regularly earn from $1,900 to $2,000 a year. They are
    paid time and a half for overtime. We made this inquiry into overtime
    work, to see whether fatigue during overtime tended to increase accidents;
    but we did not find a relation between the two. Even in excessive overtime,
    indicated by average annual earnings of at least $2,300, we found the
    same proportion among the high and the low accident groups. Perhaps the
    explanation is that the men who secure this extra work are older, safer
    men, as assignment to overtime work is strictly in order of seniority,
    among the men available.)

first group and 9 per cent of the second had some health defect as determined by the medical examinations.

At least one of these three conditions was present in 77 per cent of the high-accident cases, and in 23 per cent of the low-accident cases. The most interesting finding, however, comes to light when we compare the proportions of high-accident and low-accident men among the thirty-eight motormen who had more than one of these conditions. Thirty-six of them are high-accident men; only two are low. Stated differently, the presence of more than one of these conditions in the same individual occurs in 42 per cent of the high group, but in only 2 per cent of the low group. In other words, within this sample of eighty-six motormen, if a man is defective in health and aptitude, or health and personality, or aptitude and personality, as measured in the way here described, the chances are less than one in twenty that he will be found in the low-accident group.

Findings of this kind emphasize the necessity of individualized treatment of accident proneness, a treatment based on an intelligent diagnosis of behavior patterns exhibited both on and off the job, supplemented when necessary by medical and psychological examinations. Treatment then takes the form of specific instruction, medical attention, encouragement, or discipline, as the individual case may require, together with close supervisory follow-up. Much can be accomplished through the regular channels of supervision, when street inspectors, safety supervisors, and division superintendents all accept their share of responsibility for dealing with the motormen and bus operators in a manner which clearly recognizes the differences between the individuals whose work they oversee—differences in ability and strength, differences in skill and alertness, differences in temperament and health, differences in physique and age, differences in knowledge and training, differences in family circumstances and financial status, differences in worries, anxieties, hopes, and ambitions.

Not only in transportation companies have these basic principles and practical methods of eliminating accidents proved valid. They apply equally to safety work in factories, on aviation fields, and on the public highway.

Understanding and Control of Fatigue. The effects of excessive work—whether traceable to long hours, rapid pace, insufficient rests, muscular overload, nervous tension, inefficient ways of working, lack of training, or other cause—are sometimes primarily physiological, sometimes psychological.

Decreased capacity to do the work in hand is an obvious indicator of changes that work produces. By some investigators, reduced output has been taken as a measure of fatigue. Others have resorted to counts of accidents, errors, or spoiled work as indexes. Such studies have had to reckon with the amazing power of the human being to spurt, to nerve himself to special effort, to pull himself together and perform his task with precision and speed in spite of weariness or exhaustion. The problem is complicated—particularly so since there has been found no close relationship between the objective facts of decreased capacity for work and the subjective feelings of being tired.

Feelings of weariness are often entangled with feelings of boredom, ennui, or distaste for the job. So investigators have sometimes found themselves facing questions about the nature of monotony, or interest in work as related to differences of ability and personality, when their initial problem was that of the nature, causes, or effects of fatigue.

Indeed, the concept of industrial fatigue has proved to be rather ambiguous. Richard M. Page, for example, favors abandoning the term and speaking instead of "energy-cost," a quantity the physiologists are in a way to measure with a good deal of precision. It is of utmost importance to know the energy-cost of different kinds of work performed by different individuals, under different conditions. The feeling-cost, or the price the worker pays in terms of discomfort, pain, weariness, or distress, is every whit as real. It is of immediate concern to both management and worker, even though it cannot be measured as accurately as the amount of oxygen consumed.

To have analyzed this problem of industrial fatigue into separate components and formulated it clearly, is of itself no small accomplishment. To have forged the tools of research for recording and measuring the mental and the bodily phenomena, is another long step toward understanding and control.

Like the effects of practice and of variety of task, the effects of fatigue are of importance partly because of their interrelations with other phenomena. Psychopathologists like Elton Mayo have stressed the disintegrating effects of pessimistic revery, which often seizes a fatigued person occupied with uninteresting work, and have shown that properly systematized rests—especially when the worker has been taught a technique of relaxation—are sometimes the best cure for disgruntlement, radicalism, and excessive labor turnover. It is the needlessly tired man who most easily gets irritated at his boss and develops a grudge or foments a local insurrection. Management consequently studies both the physiology and the psychology of fatigue, with an eye to harmonious relations as well as to health and productivity.

With the widespread use of labor-saving devices and automatic machinery, however, the problem of industrial fatigue has become less serious than that of boredom. The nature of monotony, its consequences, and ways of combating it, have been studied both here and abroad, notably by Wyatt, May Smith, and Farmer, of the British

Industrial Health Research Board. Repetitive work, as such, is not necessarily productive of ennui, for monotony is inherent not in the work itself but in the task in relation to the personality of the worker. Fortunately many ways have been discovered to adjust employees to such tasks and to help them to find therein those elements of interest which add zest and satisfaction to what a casual observer might consider stultifying monotony.

Conclusion. A swiftly evolving industrial civilization such as ours needs a well-laid scientific foundation for its social engineering. Toward such a basis of knowledge about the influences determining our conduct and our feelings while at work, it has been the purpose of industrial psychology to contribute.

In this hasty glance over the field, we have noted outposts of research in factories, offices, and stores as well as in universities and government bureaus. The weapons of attack are the familiar ones of observation, measurement, statistical analysis, and controlled experiment. Among the trophies already captured, five which it has not here been possible to treat at length must, in conclusion, at least be mentioned:

1. Adult learning of specific knowledges and new skills can be profitably continued far beyond the ages at which common sense has usually called a halt.

2. While it is obviously advisable to choose an occupation most closely corresponding to natural interests and talents, nevertheless interests can be developed and many grave handicaps of native endowment compensated for, through special training.

3. Effects of temperature, ventilation, lighting, and other physical conditions under which work is done, have been measured and found to be real determiners of output and of morale, but far less potent than the personal factor of supervision. Extra financial incentives, too, have

[159]

less effect than the presence of an intelligent, competent, interested, friendly supervisor.

4. Of all the motives which lead American workers voluntarily to restrict their output, the two which operate most extensively are fear of rate cutting, and fear of layoff.

5. Scientifically sound and feasible methods of investigation in industrial psychology have been developed and validated, so that many pressing problems of vocational adjustment may now be entrusted to it with confidence.

Investigations such as those which have already been made in this field, devoted to the understanding and improvement of human adjustments in the working situation, are of benefit to both employer and employee, as two of my European colleagues quite unwittingly showed. After attending the Fourth International Conference for Technopsychology in Paris in 1927, it was my privilege to visit one of the leading industrial psychologists of Switzerland and to observe his work in the Suchard chocolate factory in Neuchâtel. I asked him how he defined the field of industrial psychology. "What do you, as psychologist, undertake to do in this factory?"

"The aim of the industrial psychologist," he said, "is to see that the workers leave the plant at night without being fatigued, irritated, or nervous."

"Good," I replied; "but just what do you do in order to bring this about?"

"First, I study the workers at their work to see in what ways it is possible to rearrange the layout, simplify the movements required, and so make it possible for them to do more work with less expenditure of energy. Secondly, I look to the training of the workers in order that they may all be taught how to do their work in the best and easiest ways. Third, in cooperation with the employment department, I have used tests of various abilities, to help in placing employees in work most closely in line with their natural aptitudes."

"Are you not concerned also with questions of wages and profits?"

"Not at all," he said, emphatically; "those matters are for the economist, not the psychologist." Even when I urged that these economic considerations had a profound effect on workers' feelings and efforts, he still maintained that they were quite beyond his province.

Two days later I was visiting the laboratory of a distinguished German exponent of industrial psychology, with whom, many years before, I had been a fellow student in Berlin. In those days there was only one book of importance in the library of industrial psychology, namely, Münsterberg's pioneer work, *Grundzüge der Psychotechnik*. I asked my German friend the same questions I had put to the Swiss psychologist. "What do you, as an industrial psychologist, actually do? Are you at all interested in the economic aspects of industrial work?"

"I never touch anything," he said, "which does not mean profit to the employer."

"That sounds like shrewd business practice," I replied. "But tell me, just what do you do in order to increase profits?"

"You see these precise recording mechanisms," he said, "which I take into the rolling mills and factories to use in studying the workers at their work, in order to find out how they can accomplish more with less output of energy. I find it necessary then to instruct the foremen, to improve their training methods so that all the workers may be taught the better ways of doing their work. And finally, you see all these psychological tests which I have perfected. They are invaluable as aids to placement of employees in those lines of work which most closely fit their abilities and natural aptitudes."

Approaching the industrial worker from opposite angles, these two scientists had nevertheless arrived at the same definition of their duties and were using similar methods. What more striking illustration can there be of the fact that in large measure the interests of employer and employee are one. Both benefit from elimination of fatigue, simplification of work, improvement of training, and correct placement of the worker. Indeed, when I asked the president of the chocolate works why he employed a psychologist whose sole concern was the interest of the employees, he instantly replied, "I find that it pays."

Such are in brief some representative accomplishments of this young science of industrial psychology. They hint at what may be expected in future years, as the ideals of scientific method and the techniques of experimental investigation come to be used more and more widely, by industrial psychologists, psychiatrists, physiologists, and management engineers, working together on problems intimately related to a large and significant sector of life.

Chapter VI

HEREDITY

by Edward M. East

DESCRIPTIONS of the properties of various chemical compounds are found in Egyptian papyri; the science of
chemistry is no older than the American commonwealth.
Exact knowledge of the composition of substances and of
the transformations which they undergo was quite impossible before the atomic hypothesis was established by the
experimental demonstration of the Law of Definite and
Multiple Proportions. With the conception of the molecule
as the unit of matter identifiable in mass and of the atom as
the elementary unit, precise information concerning the
reaction of substances under known conditions was attainable, though no one has ever seen either molecule or atom.
A similar statement may be made about heredity. Doubtless
some of our paleolithic ancestors noticed whom the new
baby resembled, and formulated theories to explain the
situation. At all events, the ancient Egyptians and Baby-
lonians must have known something about inheritance,
for they left graphic records showing highly improved
breeds of domestic animals and of cultivated plants. Yet
the sum total of previous experience up to the middle of
the nineteenth century had yielded no more penetrating
solution of the mystery than the adage "Like produces
like," a proverb which, like many another, is not true.
There is something radically wrong with such a statement
as an expression of natural law, in view of the knowledge
that two snow-white rabbits may produce litter after

[163]

litter of coal-black progeny. The saying probably became current simply to generalize on the undeniable fact that the cub of a fox is another fox, while the foal of a mare is a horse; but as a genetic principle it is lame and impotent. It would not have gained popular approval had people appreciated the logical consequences of the common observation that a child never wholly resembles one of its two parents; for the obvious deduction from this fact is that the physical basis of heredity must be a set of discrete units gathered together in the germ cells, and that the laws of heredity must be the principles which govern the distribution of these units during the maturation of the germ cells and the fertilization of the egg. When such an idea did take form in the brain of a man who could not rest until he had tested its validity, the science of genetics was born. Gregor Mendel was the man.

Mendel was a physicist by training. As such he was keenly aware of the value of dealing with the lowest possible number of variables, of controlling all experimentation carefully, and of searching for statistical relationships among the observations recorded. He chose to study inheritance in the garden pea rather than light or sound; but the choice did not make him forget the rigorous discipline of earlier days. He selected pairs of varieties to be crossed which differed but by a single striking character. He made sure that he was dealing with uniform material by self-pollinating plants of each type and studying the resulting progeny. The variety which bred true for its distinctive trait was used; the variety which did not breed true was discarded. He then made seven crosses, each involving only one pair of contrasting characters; and followed each lone character through several generations produced by self-fertilization. Previous hybridization experiments had come to naught because the varieties used as parents had differed by hundreds of characters, and the results obtained had been too complex for analysis. Mendel avoided this mistake. Only when he was convinced that he knew exactly what would happen under the simplest conditions, did he try to steer his way through the complications introduced when two or three character distinctions were under observation.

The first important fact brought to light by Mendel's experiments was that the two members of a given pair of contrasting characters may express themselves differently in the hybrids. In the pea, if a pure-breeding red-flowered plant is crossed with a pure-breeding white-flowered plant, the flowers of the hybrids are always red. Mendel called the character that was expressed in the hybrid the *dominant* trait. The other member he called *recessive*, since it had been suppressed only temporarily and appeared again in later generations. Dominance has not proved to be a universal phenomenon, however; in many crosses the hybrids are intermediate between the parents in their appearance.

It was the behavior of the second hybrid generation rather than the original cross that gave Mendel the clue to the first real law of inheritance, now known as the Law of Segregation. In this generation dominants and recessives appeared in ratios which always approached the limiting proportion three dominant to one recessive. These recessives always bred true. The dominants, on the other hand, could be divided into two types; one-third bred true, while two-thirds behaved as did the first hybrid generation.

It is plain that such results could be obtained if half of the reproductive cells of each sex contained unit factors, or genes, for the dominant character, and half contained genes for the recessive character, provided the union of these cells was a matter of chance. And such was Mendel's interpretation. It involved several radically new conceptions. The body cells of an organism were assumed to possess two sets of the units of inheritance, one from the father and one from the mother; and in pure-breeding types each set was assumed to be alike. The reproductive cells, naturally, were supposed to contain only one set of genes, which functioned as if they were wholly uninfluenced by association with the companions forming the other set, during the succession of cell divisions previous to the maturation of the germ cells. This being the case, when a germ cell containing a dominant gene unites with one containing a recessive gene, the body cells of the hybrid will each contain a gene D and a gene R; but when the germ cells of the hybrid are formed, the dominant gene D will pass into one, while the recessive gene R will pass into another.

The second principle of inheritance discovered through the pea experiments is known as the Law of Independent Assortment. It states that the behavior of each homologous pair of genes in a hybrid, and presumably of each homologous pair of genes in a pure stock also, is independent of the behavior of all other pairs of genes.

The experiment which led to the enunciation of this law was a cross between a pea plant having round and yellow seeds with one having wrinkled and green seeds. The hybrid individuals were round and yellow, for these were the dominant characteristics. But the progeny of the hybrids were not solely of the grandparental types. Instead, the two contrasting pairs of characters appeared in all of the combinations possible. There were round and yellow peas, round and green peas, wrinkled and yellow peas, and wrinkled and green peas; and the proportions of these types invariably approached the ratio 9:3:3:1. Plainly such a result could only be obtained if the germ cells of the hybrids had received one member of each pair of genes by independent segregation. This would mean, if the two genetic differences be represented by *Aa* and *Bb*, that the

[166]

four types of germ cells *AB*, *Ab*, *aB*, and *ab* were formed in equal quantities. If one multiplies egg cells *AB* + *Ab* + *aB* + *ab* by sperm cells *AB* + *Ab* + *aB* + *ab*, having regard for the phenomenon of dominance, plants duplex in genetic constitution are obtained in the 9:3:3:1 ratio. This was the situation actually found.

Numerous experiments modeled after those of Mendel were carried out during the early years of the twentieth century; and gradually more and more complex cases of inheritance yielded to factorial analysis. All possible modifications of the fundamental di-hybrid ratio (also tri-hybrid and higher ratios) of 9:3:3:1 were discovered. These modifications were due to the fact that certain types of genes produced results which made two or more of the members of this ratio resemble each other superficially, though their breeding powers were quite different. Thus hybrids produced populations which could be divided into distinct groups in such ratios as 13:3, 9:7, and 9:3:4; yet continued breeding tests proved that each was simply a modified 9:3:3:1 ratio.

Perhaps the most important result of this early work was the replacement of the old notion that *like produces like* by the idea that *each gene produces genes like itself*. Exact analysis of pedigree culture data had shown that individuals which could not be distinguished from each other by any ordinary test produced progenies which did not resemble each other. Part of this interesting situation was cleared up by the recognition of the important rôle played by dominance, since *AABB* individuals would look like *AaBB*, *AABb*, and *AaBb* individuals, though each type would behave differently in reproduction. But this was not the whole story. It was found that genes *AA* could be the determinative factors in the production of characters that would hide the characters produced by genes *BB*. In such cases individuals containing the genes *AA* (or *Aa*) would be alike in appear-

ance whether they contained genes *BB* (or *Bb*) or not.
Moreover, it was discovered that sometimes two genes, let
us say *AA* and *BB*, had to be present together for the production of a detectable character. Thus a white sweet
pea *AAbb*, crossed with another white sweet pea *aaBB*,
produces a purple hybrid *AaBb*, since both the *A* gene and
the *B* gene are required for the manufacture of the purple
pigment. This discovery, by the way, cleared up a hoary
old biological puzzle—namely, why certain crosses show
reversion to the type characteristic of some bygone
ancestor.

The numerous minor discoveries made during this period
need not concern us here, since the objective of the first
part of this essay is the groundwork of present-day genetic
philosophy. It is necessary, however, to call attention to
the general concept of gene behavior which slowly took
form in the minds of genetic investigators. From the first
it did not seem probable that even such an apparently
unimportant gene as, for example, the one which differentiates brown eyes from blue in the human race, could
have only a single function. Were this true, an organism
would be merely a genetic mosaic, a piece of animated
tiling. And this proved not to be the case. Genes are units
in inheritance, but are not units in development. Development is a coordinated affair which one may think of as the
progressive unfolding of the organism as a whole. In this
unfolding a single gene has manifold duties, and numerous
genes contribute their quotas toward the fulfillment of a
seemingly simple task. For instance, one must assume that
the mammalian eye is the result of the combined activity
of hundreds of genes; one must also assume that each of
these genes, though primarily concerned with eye development, may have minor effects on many other organs.

It took only five or six years from the beginning of the
genetic renaissance, which came with the rediscovery in

1900 of Mendel's thirty-five-year-old paper, for biologists to realize that the Mendelian type of inheritance derived from some fundamental mechanism. This conclusion was justified because Mendelian inheritance had been found in various orders of flowering plants and, among animals, in mollusks, insects, fishes, birds, and mammals. Man himself was no exception. Carefully collected genealogies had already shown that a dozen or more contrasting conditions such as sound mentality and feeble-mindedness, ordinary stature and dwarfness, and normal vision and color-blindness were due to differences in a single gene.

In spite of the facts available at this time, the criticism was often raised that this interpretation could not serve for all forms of inheritance, even in a given species, because quantitative characters—that is to say, size characters—could not possibly be inherited in this manner. These objections stimulated work upon size inheritance; and shortly the Theory of Multiple Factors, a theory which brought the inheritance of all classes of characters under one interpretation, was announced. It consisted merely in the assumption that genes Aa, Bb, Cc, etc., can exert cumulative effects directed either toward the development of the whole body or toward that of a single organ. Because of the complexity of such inheritance, and more especially because of the extraordinary effects which slight differences in environmental factors have upon size characters, this theory was difficult to prove. But by analyzing statistically the results from ordinary Mendelian inheritance, and then using the statistical procedure to analyze experimental data, satisfactory proofs were offered.

The Mendelian interpretation of heredity having been shown to be satisfactory for all characteristics in widely separated groups of animals and plants, a fundamental underlying mechanism, common to these groups, was sought which would provide the gene distribution required

[169]

by the experimental data. It was found in the behavior of the chromosomes during sexual reproduction.

The essential feature of sexual reproduction is the formation of a new organism by the union of two cells, the egg and the sperm. Now the egg is usually very much larger than the sperm, yet the tiny sperm is just as effective in transmitting genes to the offspring as the huge egg. The physical basis of inheritance, therefore, must lie in some cell component present in equal quantity in both germ cells. That component is the chromatin. The chromatin undergoes numerous changes during the life history of a cell, but at some point in the process it always fragments into bodies known as chromosomes, having a definite number, shape, and size for each species. There are twenty in corn, forty-two in certain wheats, forty-four in rabbits, forty-eight in man.

The chromosomes are the chief performers in ordinary cell division (mitosis). They initiate the process; and they alone, among the constituents of the cell, go through an exact dichotomy. Probably they also have a good deal to do with directing growth and differentiation; though what their precise duties are in this connection, is still questionable. But their really star rôle is associated with reproduction. During the preparation of the germ cells for fertilization (meiosis), their behavior is precisely what is required by the data from all crossing experiments for carriers of the genes.

As the cells of every organism are characterized by a definite number of these bodies, and as every new individual is ordinarily produced by the union of two cells, it follows that some provision must be made for reducing chromosome numbers prior to fertilization. What happens is virtually this: when either type of germ cell matures, the chromosomes line up in pairs, and one member of each pair passes to a daughter cell. Thus, if there are two pairs of chromosomes which we may designate by the letters A and a and B and b, the resulting daughter cells may be AB and ab, or they may be Ab and aB.

It is not difficult to see that this arrangement provides the physical basis of heredity which the pedigree culture investigations found to be necessary. It must be assumed that each chromosome carries numerous genes which act as determiners of the various traits which will characterize the mature individual. Each organism possesses two complete sets of these determiners because it possesses two complete sets of chromosomes. We know that each chromosome set is complete because certain plants and animals, with proper treatment, can be induced to develop when but one set is present. One set comes from the father and the other set from the mother, as can be demonstrated by crossing two species where the chromosomes of each set differ in size. At maturity, this creature will produce either eggs or sperms, and its germ cells will possess but one complete set of chromosomes. Any given egg or sperm will have one representative of the first pair of chromosomes, and it will be a matter of mere chance whether it came from the father or from the mother; similarly, it will have one representative of the second pair of chromosomes, which also may have originated with either the father or the mother; and so on throughout the whole series. There is, then, an orderly means by which the genes from the two parents pass to the germ cells which are to function in producing the new generation; and, generally speaking, this process is one by which any given germ cell receives one, and only one, gene-packed chromosome from pairs of such chromosomes which have come from the maternal and from the paternal side of the house.

Obviously with such a mechanism for passing genes from one generation to another, the number of chromosome pairs possessed by the species in question must represent

[171]

the number of independent assortments of gene packets which can be made, since the genes carried by a single chromosome would tend to be inherited together. Experimentation has shown that such a principle actually does exist. It is known as the Limitation of the Linkage Groups. If n chromosomes characterize a species, then only 2^n different germ cells can be formed through the choice of chromosomes at the reduction division.

The chromosomes do indeed pass as entities to the daughter cells at the reduction division, which is the essential feature of germ cell maturation; yet this is not the only way that genes can be distributed. If it were, we should have to consider the whole chromosome as the unit of heredity. Actually, any given chromosome contains a packet of genes which tend to be inherited together; but individual members from this packet may be exchanged for homologous genes from the mate of this chromosome at the period when like pairs of chromosomes come together, just before the reduction division occurs. This phenomenon is known as *crossing-over;* and by studying its peculiarities intensively, an extraordinarily precise conception of the architecture of the germ plasm has been obtained.

Individual chromosomes can be tagged by the possession of particular genes, and their behavior followed in hybridization experiments. Such investigations have shown that the genes are placed in the chromosomes in a linear order, like beads on a string. The Linear Order of the Genes is, in fact, one of the basic tenets of genetics. Moreover, when crossing-over occurs, the genes ordinarily recombine in the same linear order. If, for example, a pair of homologous chromosomes contain genes *ABCDEFGHI* and *abcdefghi*, then a cross-over may occur which will result in chromosomes having the genes lined up as *ABCdefghi* and *abcDEFGHI*; but the order of the genes is usually retained. If any change in order does occur, it is such that the whole

[172]

piece of chromosome which has been exchanged with the piece belonging originally to its homologue is inverted. In other words, the chromosome may sometimes have the genes lined up as *ABCihgfed*. When such a thing happens, it is known as a chromosome mutation, since it usually produces a change in the traits exhibited by the organism.

Other chromosome mutations also occur in rare instances, and each usually produces a detectable change in the individual. A whole extra chromosome may be received by a germ cell owing to the failure of a particular pair to divide. This occurrence naturally leaves one germ cell with $n - 1$ chromosomes. In such cases the deficient germ cell usually does not function. The redundant germ cell may occasionally mate with a normal germ cell, however, with the result that an individual is produced having $2n + 1$ chromosomes. Other chromosome mutations have also been traced, and their study has yielded significant results. Pieces of chromosomes are lost; unequal crossing-over occurs; and there are duplications and translocations of chromosome parts. Following such instances in pedigree cultures by means of known genes has established the Chromosome Theory of Heredity beyond all shadow of doubt, and has thrown new light on the way in which evolution has taken place.

This brief glimpse of one of the main phases of genetic work gives fair support[1] for the statement that the me-

[1] Detailed proof of the genetic principles mentioned (and many others unmentioned), together with the consequences which flow from them, can be found in numerous elementary textbooks. "The Principles of Genetics," by E. W. Sinnott and L. C. Dunn (McGraw-Hill Book Company, Inc., New York, 1925), and "Genetics in Relation to Agriculture," by E. B. Babcock and R. E. Clausen (McGraw-Hill Book Company, Inc., New York, 2^d ed., 1927), are recommended. For information on the cellular basis of genetics, the reader should refer to "The Cell in Development and Heredity," by E. B. Wilson (The Macmillan Company, New York, 3^d ed., 1925). For the effect of modern genetic doctrines on the evolution concept, he should see "A Critique of the Theory of Evolution," by T.

chanical problem of heredity has been solved as definitely as the mechanical problem of molecular structure. Let us recapitulate. Every organism, whether man or beast, whether flowering plant or fern, starts life as a combination of more or less independently inherited genes. The genes are the units of heredity, the physical connection between one generation and its successor, and are the basis of all developmental changes. They are self-perpetuating bodies which grow and divide throughout long periods of racial history without exhibiting discernible variation in the functions they perform; yet in rare cases any given gene may take on a new constitution and thus, if a functional germ cell results, provide new raw material for the sieve of natural selection. The gene pattern of every individual is definite and distinctive for that individual, and is ordinarily made up of two complete sets of genes, one contributed by the egg and one by the sperm. Each mature egg or sperm possesses one complete set of genes, chosen from these two sets by means of a regularly constituted method of gene distribution. Normally, the essential features of this distribution are as follows: Any homologous pair of chromosomes may exchange one or more genes, provided the exchange is equable and is carried out in obedience to regulations. The entire packet of genes of each respective pair of chromosomes is then passed to the two daughter cells which become the mature reproductive cells—through a second equatorial division having no known genetic significance—by the free assortment of one member of each pair.

The emphasis which has been placed upon gene patterns should not be construed as meaning that no other factors have determinative value in the final result that is called

H. Morgan (Princeton University Press, Princeton, 1916), or "The Genetical Theory of Natural Selection," by R. A. Fisher (The Clarendon Press, Oxford, 1930).

the organism. Such an idea would be incorrect. The fertilized egg has a gene pattern, it is true, and the type of gene pattern which it possesses has a governing influence on the characteristics of the adult; but the fertilized egg also contains cytoplasm, and cytoplasmic activities are by no means to be neglected. Further, a fertilized egg must meet hospitable conditions in order to develop; and these conditions may be varied. And finally, the behavior of the individual which results from the interaction of gene pattern, cytoplasm, and environment, is also conditioned by experience.

Cytoplasm is the term used for the mass of protoplasm in which the nucleus containing the chromosomes is embedded. Its precise function is not yet wholly clear. Perhaps we should not go far wrong if we considered it to be simply one of the internal factors of environment. It appears to contain no essential substances which, like the chromosomes, must be transferred to the daughter cells at each cell division in order to have the machine function effectively. On the contrary, the cytoplasm seems to be built up anew each time a cell divides, as if it were being manufactured under the direction of the nucleus. Nevertheless, it plays an extraordinary part in early development. In certain animals it can be shown that the cytoplasm of the unfertilized egg gradually arranges itself into three strata, and that these strata constitute the groundwork of the new individual. For example, in the sea urchin one layer becomes the body covering, one the inner part of the alimentary tract, and one the skeleton; and if any layer is disturbed, there is a corresponding disturbance in these particular parts. Now this cytoplasmic behavior may be directed by the chromosomes, and there is some evidence that this is so; but at all events, whatever structural differentiation takes place in the *unfertilized* egg occurs under the sole influence of the genes of the mother. The

[175]

general ground plan of the body, perhaps the common characteristics of the order or the family, is due to the mother alone. Another plausible conclusion can also be drawn from these facts. Since if one of the cytoplasmic layers of the sea urchin is removed by a tiny pipette, the part of the body to which it should give rise does not develop, it appears to follow that the nucleus can perform its governmental function but once in a given cell generation.

Another important group of internal environmental factors may be termed factors of position. With favorable material, such as the sea urchin, the two cells resulting from the first division of the fertilized egg (or even the four-celled and eight-celled stages of later divisions) may be shaken apart. Each cell then produces a normal surface layer over the area where it has previously been in contact, and restores the respiratory conditions necessary to, and characteristic of, the normal egg. The two cells, which would have produced the right and left halves of a single individual, now proceed to develop into complete organisms. With other material, separation of the cells results in the production of right- and left-half organisms. With still other material—the frog is a good example— the egg itself may be cut into two parts; and each part will produce a half individual. What is the cause of these peculiar results? The answer given by the biologists best able to judge is this: Organisms vary somewhat in the detailed maneuvers by which they differentiate; but in every case development is a process of sorting out the various cytoplasmic materials and is directed by the genes under certain restrictions offered by cell position.

The conclusion that the course of development taken by any cell lineage depends upon its relations to other cells is not a speculation or an indirect inference; it is a direct conclusion based on experimental evidence. In the early stages of embryonic development, cells which would have produced parts of the central nervous system will produce skin if transplanted to a region that would normally produce skin. An organization center exists at a given place in the hollow ball of cells comprising the gastrula, and what each cell becomes depends upon its position with reference to this center. The organization center itself can be shifted experimentally, and, following the change, the whole organization pattern of the embryo is altered.

In later stages of development this formative power is lost. Differentiation having passed a certain point, the fate of any given structure is more firmly fixed. Liver cells, or spleen cells, or brain cells, or skin cells remain the cells of these types, no matter where they may be placed. Yet an opportunity for a limited degree of variable development still exists in individuals having identical gene patterns.

In the first place, the developmental machinery is not perfect. Slips may occur; and from these slips some diversity of structure and function may ensue. These workshop errors may affect organs having little influence on general physiological activity, and the result be negligible; but if they retard or accelerate the energy of such essential organs as the ductless glands, the consequences may be more serious. In the second place, an alteration in the physical environment may change the rates at which chemical changes take place, and thus entail poor development or superior development, as the case may be; but the changes induced by such causes are quantitative rather than qualitative. In fact, one should not expect to have the normal development pattern upset markedly by either of these causes. An organism is a dainty piece of machinery; and the higher it is in the evolutionary scheme, the more delicate it turns out to be. If slips in development occur, therefore, or if inhospitable conditions are met, growth is more likely to stop altogether than to continue in an abnormal manner.

These general considerations give us a clearer idea of the processes collectively known as growth; but, after all, they do not answer the question of greatest interest. We want to know precisely what variations can be expected—due to the deviations in environmental factors that are ordinarily confronted during development—in a series of individuals having identical gene patterns. We may admit that the gene equipment makes it possible for an egg to develop into a potato, or a fish, or a man, as the case may be, with the appropriate characteristics which make these organisms distinguishable from one another. We may admit that the genetic constitution even marks out rather definitely what kind of potato or fish or man shall develop. But what are the extreme developmental variations attributable to the environment as a whole?

It is difficult to give an answer that will cover all cases, for various classes of organisms have different powers of adjustment and different factors of sensitivity. The type of plant or animal concerned must therefore be taken into consideration. In general, capacity for adjustment diminishes and sensitivity increases as we go up in the evolutionary scale. Thus the lower orders of plants are less sensitive than the higher orders; aquatic animals are less impressionable than land animals; cold-blooded animals are less responsive than warm-blooded animals. The mammals, and in particular man, are extremely sensitive. These facts give rise to an interesting situation. Experimental alteration of the environment can cause greater effects on the lower organisms than on the higher organisms simply because the latter have developed natural devices which preclude any effective modification of conditions. It is not to be assumed, however, that even the lower organisms have unlimited powers of response. They are able to undergo only such changes as are permitted by the gene pattern; and these changes are limited in type and degree. An experimenter can modify the chemical constitution and the temperature of sea water quite materially and thus obtain bizarre fishes having Cyclopean eyes or two heads or divided tails; but these monstrosities are not chaotic expressions of change in every direction; on the contrary, they exhibit the natural potentialities of the gene patterns under the novel conditions presented. We need not dwell, however, on the character, degree, or frequency of the variations produced in such special cases. The matter to be emphasized, it seems to me, is that such animals as the higher mammals cannot have their environmental conditions altered materially and live. Just consider for a moment the extreme nicety of man's regulatory apparatus. Temperature is controlled; moisture is regulated; and the types of food permitted are severely restricted. If his temperature rises half a degree, he is ill. If he does not get a balanced ration consisting only of fats, carbohydrates, proteids, essential minerals and vitamines, growth is prohibited. He cannot even synthesize his own protoplasm unless he has all the necessary amino-acids. And if his blood absorbs a tiny fraction of any one of a million different chemicals, he dies. One should not expect to find great plasticity under such conditions. Moreover, man is a delicate machine in another sense. When developmental errors occur and monstrosities are formed, these deviants are seldom viable.

Generally speaking, therefore, the effect of any ordinary change of environment is negligible. It is usually confined to the production of well-nourished or ill-nourished offspring; and this is true, whether one is speaking of plants or animals, of unspecialized forms or highly specialized forms. But it should not be forgotten that even in man himself an occasional serious aberration is attributable to the environment. An accident may prevent normal maturation of the central nervous system, resulting in imbecility or feeble-mindedness; absence of iodine in the diet may affect the growth of the thyroid gland, resulting in a condition resembling cretinism. These inductions are supported by thousands of experiments in which the genetic constitution has been held constant, while the environment has been varied; they are supported by thousands of other experiments in which the genetic constitution has been varied, while the environment has been held constant.

It is not wholly fair to attribute these profound conclusions to the geneticist, however; the farmer discovered the facts years ago, while the geneticist was floundering in a sea of philosophic doubts. Do not, I beg of you, ask a farmer whether he can turn a hackney colt into a thoroughbred by giving it special food and training, if you wish to retain a reputation for intelligence. Do not ask him what changes can be made in a Concord grape or a Winesap apple or an Elberta peach by an altered soil or a diverse climate. He knows well enough that every plant of each of these asexually propagated varieties has the same genetic constitution as all the others, and that, for this reason, they are recognizable members of their respective groups wherever they are grown.

The facts being what they are, therefore, one should not ask whether heredity or environment is the more powerful. It is a silly question. Both are essential. They are collaborating artists, their finished product the individual, yet with different rôles to play as moulders of destiny. The gene pattern received by the fertilized egg definitely delimits the end result. One does not gather figs from thistles. Nevertheless, environments can differ to some extent without preventing development; and these diverse factors may have a recognizable effect upon the final product. Ordinarily such effects are restricted to such quantitative changes as we are familiar with under conditions of good

[180]

hygiene and bad hygiene. And the alterations to be expected in man by environmental diversity are probably less than for any other organism, owing to the delicacy of his organization and the mechanisms that have been produced to keep his environment constant.

I know of no better analogy to use in comparing the respective functions of heredity and environment than one borrowed from photography. The gene plan of the fertilized egg makes it like an exposed plate. The potentiality of a picture is there, waiting to be developed. The environment is the developer. It can make or mar the picture, but that picture will have the same general character in any event. Translating these matters into the terms of sociology, one may say: Give the growing child the best conditions possible; it is highly desirable; but do not expect to change the character of his features or the quality of his brains.

If it has not been made perfectly clear before, the last sentence will serve to indicate that man cannot be excluded from the aggregation of organisms ruled by genetic law. Well, why should he be excluded? Man is simply a higher mammal, with a structure and a set of physiological processes very similar to those of other mammals high in the evolutionary series. From these facts one might assume that the method by which his heritage is passed on would be essentially the same as that which is common to all the remaining members of the animal and vegetable kingdoms which reproduce sexually. But no such abstract deduction is necessary. The observations of competent cytologists have shown conclusively that human body cells contain forty-eight chromosomes, and that, at the maturation of the germ cells, these chromosomes undergo the special type of reduction division required for the distribution of the genes by the Chromosome Theory of Heredity. In addition, analyses of voluminous genealogical records have demonstrated that numerous character differences, affecting each

[181]

of the various organ systems, are inherited as we should expect them to be inherited if they are due to one or more genes carried by the chromosomes and distributed by the normal mechanism for such distribution. Sex is even fixed by the possession of a particular type of chromosome carried by all of the eggs but by only one-half of the sperms. And it has been proved that hemophilia (loss of the blood-clotting mechanism) and color-blindness (diminished sensitivity of the cones of the eye) are carried in this same chromosome, just as sex-linked characters in other organisms are due to genes carried in this so-called X-chromosome which has the most direct influence over sex. Other more complex characters follow the same scheme. The details of gene interaction are more difficult to determine in such cases, as, for instance, skin color in negro-white crosses; yet it is significant that where complex character differences are under consideration, the genealogical records and the records from the genetics laboratories are virtually identical in character. Mathematical analysis of controlled experiments brings out certain conditions which ought to be satisfied in the genealogical records if the heredity mechanisms are comparable in the two cases; and these conditions are satisfied. Geneticists are convinced, therefore, that inheritance in man is the same as in melons, or mosquitoes, or monkeys.

Certain critics who accept this genetic generalization for physical characteristics question whether it holds for mental characteristics. And, unfortunately, the less they have studied the subject, the more positive they are in their assertions. Dr. John B. Watson, the extremist leader, completely oblivious to genetic facts, even boasts that he can make a fool or a genius of any normal child if he begins his instruction at a sufficiently early period. Since man is chiefly interested in man, it may be well to develop the genetic beliefs on this point in a little more detail. The argument will give a better orientation for the remarks on the applicability of genetic philosophy to sociology, to which the concluding pages of the chapter are devoted.

The convictions of the radical behaviorists are wholly emotional in character. For some unknown reason, they wish to believe in the innate intellectual parity of all members of the human race. Their minds are firmly *conditioned* in this respect, to use one of their own terms; hence, a knowledge of the genetic facts, if one grants such enlightenment, has little influence. Their sole argument, stripped of all verbiage, is an example of a paradox in logic that would have given William de Morgan great pleasure. They grant that the human race exhibits heritable differences in all sorts of external features, and in various parts of the skeletal, muscular, circulatory, digestive, and secretory systems. They admit that the central nervous system is the physical basis of mental capacity and that heritable defects in it may cause feeble-mindedness. And despite these admissions, they stolidly maintain that the normal central nervous system does not vary. At least their arguments are all reducible to this simple and obvious fallacy.

The least illogical attempt to justify the position of the absolute environmentalist is the comment that since the tremendous range in intellectual attainments shown by different individuals is not paralleled by detectable physical variations, the environment must have more to do with mental than with physical development. But this argument also is fallacious. One may also say that training brings out innate differences in mentality and thus offers a method of detecting the physical differences.

It is regrettable that so many social workers, physicians, and psychologists should tend to minimize the rôle which heredity plays in human affairs merely because they jump to the erroneous conclusion that if this rôle is trivial, their own vocations are of more importance. On the contrary, the business of alleviating sorrow, sickness, and mental maladjustment is as momentous from the genetic point of view as from the environmentalist point of view. But even if this were not so, we should still have to face the facts. And the facts are roughly these. Individuals, because of their diverse gene patterns, exhibit marked differences in aptitudes, temperaments, and all other varieties of mental capacity. They have this groundwork laid down at the very beginning of development; and it is altogether likely that, owing to the rigidity of the environment entailed by mammalian physiology, there is great constancy in the unfolding processes which are carried on during the more plastic period of the life history. At puberty, or a little after, growth is finished; and the mind, as well as the body, has reached the state of complete organization known as maturity. But this does not mean that either the mind or the body becomes rigid and inflexible. It may very well be, as Watson claims, that, in a certain sense, the very young child has no mind. It has merely a rather empty brain which can be filled and its behavior conditioned by what fills it. But the brains of children, like clay pots, are of various sizes and qualities. What they will hold, and how much, is largely determined by the genic materials which went into them.

Though to those who are familiar with modern genetic data it is as redundant as two tails to a dog, direct evidence exists on this very point. It is found in various statistical studies of eminent men and their relatives, in studies of identical twins, and in studies of intelligence by means of psychological tests.

Among the extensive studies of eminent men and their relatives are those of Galton, De Candolle, Ellis, Cattell, and Woods. Naturally, such studies are full of pitfalls. It is difficult to select a series of names for study in a wholly objective manner; it is impossible to make accurate allowances for differences in opportunity; and eminence is certainly not a precise measure of intelligence. Yet, in spite of their deficiencies, these analyses have yielded some very significant results. Each investigator finds that the relatives of great men have a very much greater chance of becoming eminent than people selected at random from the general population. The difference is too great to be accounted for by any assumed difference in opportunity. Moreover, it is to be noted that *the chance of becoming great varies directly with the closeness of genetic relationship*. One must agree with Cattell, of course, that environment holds a veto over mental development, but one cannot avoid the conclusion that environment merely harnesses the forces of heredity.

Such investigations go far toward establishing Galton's important generalization that no man is likely to attain eminence unless he possess innate ability of a high order. They also support his further conclusion, though perhaps they do not establish it, that "few who possess these very high abilities can fail in achieving eminence." In other words, it is hard to keep a good man down. Or woman either. An instance is Helen Keller. Cattell once said that if Isaac Newton had been born among Hottentots, he would have announced no Laws of Motion. True enough. But he would have been the intellectual leader of the tribe. A few Hottentot Isaac Newtons, and the group would have had a different cultural history.

The evidence from identical twins points in the same direction. Identical twins come from a single fertilized egg. Each member of the pair, therefore, has the same heredity. And such twins exhibit mental attributes as well as physical characteristics which are much more nearly alike than are the traits exhibited by other children of the same family. They even tend to show this same similarity when they are separated at an early age and reared under different conditions, though the facts are not altogether harmonious in the few cases of this type that have been adequately studied.

The intelligence tests add to this evidence. Intelligence tests measure acquirements, it is true, and they have been criticized severely on this score. But the critics should realize that there is no more reason why properly safeguarded measures of intellectual acquirements should not be interpreted in terms of genes than that developed physical characters should not be thus interpreted. Two ten-year-old children, for example, show significant differences in their intelligence quotients. They are given the same studies and are tested again at the end of four years. They are now further apart mentally than they were in the earlier test. The more intelligent has advanced further than the less intelligent. What conclusion can be drawn other than that they differed in genetic constitution?

From the sum total of all the evidence geneticists have concluded that mentality is a complex trait inherited like any other complex character. Some of the numerous genes concerned presumably have varied into what one may call plus and minus types; and it takes only 20 of such genic differences, inherited independently, to give the possibility of 1,000,000 different recombinations. "If the Thinker requires 20 plus genes and the Simpleton 20 minus genes, then the Average Man may be supposed to have about 10 plus and 10 minus genes. If a family stock, by selective matings, gathers together a preponderant proportion of plus genes, its average worth will rise; conversely, if a family puts its efforts into accumulating minus genes, its social value will drop. But even the mediocrities may produce Thinkers—or Simpletons—if the constitutions of the fusing germ cells are such as to bring together the required genetic complex. Thus there is no difficulty in accounting for emergent individuals like Carlyle and

[186]

Abraham Lincoln in otherwise undistinguished families. The proportion of eminent persons in such families will be low because of great differences in quality among the parental germ cells. On a percentage basis, selected high-grade families will often produce ten, twenty, or even fifty times as many notables as mediocre families. Yet the absolute number of geniuses appearing in families of the latter type will always be high because mediocrity is plentiful."

In passing, it may be well to note that a number of papers have recently appeared in which the writers have failed to grasp the genetic concept of mentality. If it be true, these authors say, that variations in intellectual capacity can be interpreted by assuming that numerous genes combined into different patterns are involved, how is it that the contrast between a normal-minded person and an imbecile is to be represented by a mutation in a single gene? The answer is quite simple. A hundred genes, let us say, may function in producing the normal mind, and a good many of them must have varied into plus and minus types in order to account for the diverse grades of mentality actually observed. But since the evidence indicates that unions of feeble-minded persons produce only feeble-minded persons, one must conclude that only a single gene has become truly defective, though ninety-nine other genes are left to contribute to various grades of feeble-mindedness. A second defective gene may yet be discovered. Proof of its existence would be found in a union between two defectives which yielded only normal progeny, provided genetic tests of the offspring conformed with the results to be expected for individuals carrying two different genes for defectiveness. No such evidence is available at present.

Briefly, then, the genetic philosophy is this: Heredity allots to each individual certain mental and physical potentialities, governed in this allotment by the rigid laws controlling gene distribution; whether these potentialities are wholly realized or not depends upon circumstances. These facts are significant in two ways. The fundamental requirement for success in life is a satisfactory genetic constitution; this much is clear. But the old arrogant pride in blood which, at the worst, risked only a slight dilution of its powers by an ignoble union, has had to be abandoned. Dilution, whether of power or quality, is not an attribute of a gene. The geneticist yields to no one in his regard for good breeding; yet no one knows better than he the fallacies of ancestor worship. One out-cross can ruin an otherwise irreproachable lineage; one proper mating is all that is necessary to produce the true aristocrat. The genetic constitution of a distinguished family is likely to be compounded largely of good genes; hence the high probability of worth among its members; but the degenerate product of a bad genetic combination is not saved by the personal record of his ancestors. Nor, by the same token, is the genetically great to be condemned because his endowments are choice gifts from a scanty store. In the second place, a real appreciation of the great innate diversity of human beings leads to a proper conception of the tremendous importance of training. Sociologists appear to feel that education is of no consequence if this point of view is adopted. Quite the contrary. If all persons were products of the same mould, pedagogy would be a simple matter. It would merely be necessary to find out how best to polish up a single pattern. With all sorts of special aptitudes and tendencies, our instructors of youth have a more difficult task to perform.

Suppose we accept this genetic philosophy; to what practical conclusions does it drive us?

In the first place, it seems to me, must come the relinquishment of our professional acceptance of Jeffersonian democracy. Men are not created equally free or essentially equivalent. It is true, the governmental policies of the

United States have never demonstrated complete faith in Jefferson's principles in any concrete manner. Sustained and, for the most part, logical action based on the doctrine of human parity has been carried out only by the Soviet Republics. The keystone of Communism is a religious acceptance of two biological errors, to wit, that all persons have the same innate intellectual equipment, and that acquired characters are inherited. The second tenet, by the way, means to the Communist that stuffing the present generation with schooling will make their descendants more intelligent. In the United States we have not carried matters so far. But we have labored to achieve democratic perfection by assuming that suffrage can be exercised wisely by anyone reaching the age of twenty-one years, provided he or she has been properly moulded into the American pattern by a primary school education. And, Fourth of July orations to the contrary notwithstanding, this is not a well-ordered scheme.

The country has become much more socialistic than most people realize. Theoretically it is committed to a policy of free medical advice, and free mental training in the public schools. This program is partly logical and partly illogical. It is logical in that every citizen should be given the opportunity to realize on every physical and mental asset he possesses. It is illogical in that we treat all individuals as if they were Boston terriers, trimming their ears and teaching them identical tricks, so that they will fit a single given standard. It would be much better, would it not, after patching up such physical disabilities as we may, to give each child all the education he can assimilate of the type for which he is fitted? There is no point in trying to teach our twenty million morons to read and write. It is hardly worth while to prod another twenty million dullards through grammar school. It is foolish to lower high-school and college standards so that everyone who

[189]

has a mere social urge for educational gloss is able to obtain it. It would be a much better policy to reorganize the whole educational system in a way which would permit the establishment of specialized schools fitted to the different requirements of our variant population. The gifted child should be our special care. The progress of a people depends largely upon the upper 1 per cent. Dr. Cox has shown that the leaders of the world, the geniuses of various types who have made civilization, have been exceptionally intelligent as children. Intelligence tests have been sufficiently perfected to enable us to select such a group. Psychologists cannot guarantee that they will all be leaders; they can guarantee that the leaders will come from among them. What greater service to humanity could a government perform than to select and train everyone who shows promise of outstanding ability, no matter from what walk of life he comes?

If the members of each generation of United States citizens were grouped and trained according to their gifts, it might be well to give up the "one man, one vote" idea. Plural voting for the higher grades of trained intelligence is surely sensible, and it might be an incentive to accomplishment. It sounds so heretical, however, that I shall not advocate it. Yet there is no reason why one should be overcautious in suggesting educational qualifications for suffrage that are high enough to insure some real fitness in the electorate. The principle has been accepted, but as applied to-day it means nothing. The requirement in the most advanced states is merely the ability to read by rote a few sentences of the constitution. The result is a democracy of voters of which not over 50 per cent rationalize their duties. The remaining half is a mob swayed by clerical wowsers, venal editors, and political demagogues.

These suggestions may seem too generally theoretical, and therefore too vague to enlist general interest. Really they are not. They give this impression because they concern the whole political system. Perhaps the biological point of view can be made to appear more practical, however, if some definite and concrete problems where its adoption would make for a better world are mentioned.

Most readers are familiar with the fact that the larger cities are starting medical and psychiatrical clinics in connection with the courts. This is a real scientific advance in one of our most important social institutions. Our penological system is a fossilized remnant of the dark ages. One may not be convinced that law breakers come, in the main, from the feeble-minded and insane. Possibly the majority possess normal intelligence. If so, either our early teaching of the malcontents or the formulation of some of our laws is at fault. On the other hand, there is satisfactory evidence that a lot of the petty crime is committed by easily led morons, and that nearly all of our revolting crimes are committed by people who should be in psychopathic wards.

Obviously, the asocial individual is often ill and demands treatment instead of punishment. He is a human being, not a machine. There is no good reason why his case should be disposed of by the gentlemen of the bar, men versed solely in legal precedent, instead of by the psychologist and physician. And such a change in judicial procedure is imminent. But what would you say to going still further and demanding preventive medicine as well as correctives? I firmly believe that this should be done. I go so far as to advocate clinics where every child is examined both physically and mentally at stated ages.

Presumably such institutions would pay their way as public safety devices. Maniac depressives and paranoiacs would frequently be detected before they got in their deadly work. The epileptic and the feeble-minded would be earmarked, so to speak, and limited in their activities.

[191]

To-day I am told that the moronic young male naturally gravitates toward truck-driving, while the mentally defective female tends to qualify as elevator girl. Not a very wholesome situation!

But there is another and more important function which could be performed in such clinics; namely, the fitting of our various types of human pegs into the right-shaped holes. The minister, I believe, always receives a "call" to his life work. He is a favored class. All of the rest of us over thirty know what a perfectly awful time we had trying to decide upon a profession. And, having decided, we spend the rest of our lives wondering if an excellent senator or saxophone player wasn't spoiled by the decision. Perhaps we should suspect that we were fitted for higher things, no matter what the tests showed; but at least we should be in the possession of expert advice on the subject.

Other political matters where one might urge that genetic knowledge be weighed are those connected with marriage, divorce, and reproduction.

Our state marriage laws are in quite a chaotic condition. For example, three states forbid the marriage of cousins once removed, while one state permits a man the somewhat embarrassing privilege of marrying his grandmother. In few states have sufficient safeguards been placed against marriage between feeble-minded persons or persons having certain forms of insanity. Unions between genetically feeble-minded persons, it will be recalled, produce only feeble-minded children. It is clearly the province of the state, therefore, to make such unions illegal. It is also the part of wisdom to provide impediments to marriage in other cases where defective children are the probable result. How this is to be done is still a question; but the registration of certain essential data before the marriage license is granted appears to be the solution.

[192]

Our divorce laws are even worse than our marriage laws. For the most part they are projections of outworn theological tabus having ofttimes no rational basis. We are not thinking of the delight which some of our theologians appear to take in punishing a couple for a mistake in judgment by a lifetime of misery; we refer to their punishment of the next generation by refusing to see in feeble-mindedness or a heritable tendency to insanity a just cause for divorce.

There are also other laws of biological import in dire need of revision. We should humanize our statutes regarding illegitimacy, making it possible, or even inevitable, that the child born out of wedlock be legitimatized and given certain rights of protection. Our mediaeval laws regarding contraception should be modified. At present they fail to safeguard mothers against unwanted pregnancies when, through certain diseases of the heart, lungs, and kidneys, pregnancy means probable death; and they fail adequately to guarantee the child that reasonably healthy start in life which would be guaranteed by the proper spacing of children. We should also encourage the more general adoption of laws providing for the sterilization of the feeble-minded, of the type so effective in California. The California laws are sound scientifically, and they have been pronounced constitutional by the Supreme Court of the United States. "Three generations of imbeciles are enough," said Justice Holmes in his brief.

Other instances of the applicability of genetic philosophy to social problems might be given were space available. Under the circumstances I must content myself with the suggestion that it would be desirable to have intelligence tests as part of the physical examination of would-be immigrants.

The United States has been, and is, an extraordinarily prosperous nation. To what does it owe this good fortune?

I have seen plenty of answers in the public press—the type of government, the influence of religion, tariff, sound money, and so on—but none touches the mark. Given its natural resources, the pattern of a country's career is set by its small group of trained men of high intelligence, the men who deal masterfully with the problems of science, art, politics, and business. The rest of us need not fool ourselves, we are simply the chorus of this drama. But these grade A people are rare, as are also the feeble intellects which lie at the other end of the curve, the grade E people. People of grades B and D are somewhat more common; people of grade C are commonest of all.

In other words, the individuals composing our population exhibit enormous differences in innate capacity, the distribution of intelligence being such that a high-peaked symmetrical curve is formed, with mediocrity holding the center position. And it is the exceptional few at one end who mark out our path in the world. If sufficient data were available, the distribution of intelligence among the nationals of other countries would be found to be very similar in character. There might be slight differences in average capacity; such differences would be small, however, when compared with the great spread of differences within each group.

It is not difficult to see that the future progress of any nation, our own included, depends upon the proportion of people of grades A and B. If they increase, the country will prosper; if they decrease, the country will decay. There are two ways to shift the average. One may try to promote the survival of the grade A people and to retard the survival of the grade E people who are citizens of the country, which is the method ordinarily proposed by eugenicists; or one may try to promote the entrance of grade A people and to prevent the entrance of grade E people as immigrants, which is the suggestion made here.

The reader will note the avoidance of eugenic proposals in what has been said regarding problems in the sociological field. With the eugenic procedure suggested by Sir Francis Galton in 1883 everyone must have the most profound sympathy; for he defined his purpose as "the study of agencies under social control that may improve or impair the racial qualities of future generations, either physically or mentally." Even the mediaeval mind of G. K. Chesterton would have difficulty in discovering the evil character of such an ideal. But one may also confess to no wild enthusiasm for the small-knowing souls who offer social salvation in impracticable programs based on that vaguely conceived formula "the survival of the fit." The creed itself is sound enough, since social progress depends primarily upon the genetic constitution of the people of which society is composed. The irritating point in such schemes, apart from any question of feasibility, is the emotionalism so evident in the average segregation of the sheep from the goats. The *fit* would be for Bishop Cannon the Dry, for Hilaire Belloc the Catholic, for Pat Harrison the Democrat, for the Reverend Doctor Massee the Fundamentalist; they might not be so *fit* for Mencken, Joseph McCabe, Senator Vare, and Clarence Darrow. For these reasons, one is justified in feeling, regarding eugenics, much as did Charles Darwin when he wrote to Galton on the subject in virtually these words: The object is a grand one, and it is the sole feasible plan of improving the human race; yet the difficulties which stand in the way of any practical procedure are so great as to make it, I fear, Utopian.

Genetics should do more than furnish the factual foundation for projects of direct racial improvement, even though some such measures may be both sound and profitable. It has in it the makings of a broad and practical social philosophy applicable to numerous questions connected

[195]

with public health, penology, education, suffrage, and immigration—to mention only a few. The keystone of this philosophy is the proper recognition of the precise significance of genetic constitution, environment, and experience as determining factors in human behavior. I venture the opinion that the twentieth century concept of the gene as the unit of inheritance will have as much influence in directing our approach to sociological questions in a scientific manner as the nineteenth century doctrine of evolution had on the problems of technical biology.

Chapter VII

THE FRONTIERS OF MEDICINE

by Morris Fishbein

THE evolution of medicine since the time of the Edwin Smith Papyrus (2400 B.C.) is a record of gradually developing knowledge built from the contributions of thousands of workers in every nation of the world. Whenever a new discovery is made, the way is opened for innumerable contributions by men who have not, perhaps, the ingenuity or originality to develop a new conception for themselves but who may, in the laboratory or in the clinic, follow a thought or a technic set forth by another observer. Thus, the discovery of a new vaccine, serum, or antitoxin for one condition may lead to the application of similar technic for other diseases. The development of a method of making visible an internal organ or cavity, as is done with the special dye substance used in the case of the gall bladder or with lipiodol in outlining the bladder, the uterus, or the sinuses, leads the way to similar investigations in other viscera and tissues.

The invention of new devices, such as the microscope, the stethoscope, or electrocardiograph, brings about the application of the devices and technic to the study of various conditions. The announcement of the isolation of a chemotherapeutic preparation, such as salvarsan or arsphenamine, leads the way to the development of combinations of similar character applied to different diseases. The isolation of a chemical principle, as from the thyroid or the pancreas, stimulates similar work on other glands

[197]

of internal secretion. By this path medicine proceeds to conquer one field after another, so that to-day the scope of medical knowledge is far beyond the conception of any one man.

Optimists assert that eventually disease will be obliterated. They paint a glorious picture of long life with freedom from disease. Scientists know that this is a vain hope, for the human body changes, the environment changes, and diseases meet these altered conditions. Many diseases are caused by living organisms which undergo evolution exactly as man has undergone evolution. The changing habits of man affect the condition of his tissues and his health. The invention of new devices for transportation, for light, for heat, and for amusement, unquestionably affect the human body. The introduction of powerful rays may introduce new diseases due to the effects of those rays. Moreover, down through the centuries intermarriage and uncontrolled breeding have served gradually to modify the nature of the human being, so that one of the most serious phases of medical study to-day is that applied to constitution and heredity.

Specialization. Until the middle of the nineteenth century the practice of medicine was carried on by an individual who was presumed to be competent in every medical field. He could take care of a pain in the abdomen, a cough or a cold, pneumonia, tuberculosis, an eruption on the skin, infantile diarrhea, or typhoid fever. He advised for the baby, for the mother, and for the grandfather. In an emergency he would take off a limb, open the abdomen, and even the skull. Not infrequently, however, he overlooked serious conditions for the simple reason that the methods and machinery for diagnosing and treating such conditions had not been perfected.

Around 1880, the discoveries made by Pasteur definitely changed the trend of medicine. The establishment of the fact that germs cause disease, brought about control of the acute infectious diseases. It became possible to prevent their transmission from one person to another and to stop the devastating plagues which used to wipe out entire communities. Infections could be avoided through inoculation.

Associated with this knowledge came new concentration on the blood as the important medium in the human body for aiding resistance to disease and for taking care of disease conditions when they develop. Obviously, new machines had to be devised for all of these purposes, since the objects with which one deals are microscopic in size, some of them even too small to be seen by the finest microscope.

As a result of this new knowledge, medicine began to divide itself into branches, many of which were concerned with special ability to use particular instruments. The technic of employing these new devices was, in itself, an art to be acquired only by long practice. Moreover, one was concerned not only with the application of such technic, but also with a knowledge of things seen and heard by the use of the newer apparatus and an interpretation of these observations in the light of our information on disease.

Internal Medicine. The specialist in internal medicine concerns himself with disturbances affecting the heart and lungs, the circulation, the organs within the abdomen, digestive disorders, degenerative disorders, and other complications of human physiology, which are puzzling to the general practitioner. In the diagnosis of heart conditions, it is possible for any general practitioner who has kept himself up-to-date to determine whether or not diseases of the heart are present, to determine—in general— the nature of that disease, and the competency of the patient to perform certain work.

[199]

The heart is hidden away in the chest. All of its functions and the changes which it undergoes must be determined from evidence secured indirectly. It is possible by thumping the chest to outline roughly the borders of the heart and thus to know whether it is larger or smaller than it ought to be. By use of the X-ray the borders of the heart may be outlined exactly. It is possible, by listening to the heart, to determine whether or not the valves are functioning satisfactorily and to know whether it is likely that they are constricted—thus permitting an insufficient amount of blood to pass through— or are dilated—permitting blood to leak back after it has passed through. When such changes occur, characteristic murmurs arise which can be heard by means of a stethoscope.

It is possible, by listening to the beat of the heart and by feeling the pulse, to tell whether or not the beat is taking place satisfactorily, and whether or not the blood is being pushed through the blood vessels and back to the heart as it ought to be. Such measurements are, however, loose determinations. A device called the electrocardiograph measures, by photographing waves on a photographic plate, the changes in the beat and any disturbance of the mechanism of propelling the heart beat through the organ itself. Such methods are far more accurate than the use of the unaided human senses.

One may feel the pulse in the wrist with the finger, or one may attach to the wrist a device called the sphygmograph, which records the variations in a pulse beat on a moving strip of paper in such a way that the various changes may be determined. The machine is far superior to the human finger in detecting changes.

There was a time when it would have been considered equivalent to murder to push a needle into the sacs around the heart. To-day the physician frequently extracts the fluid accumulated in that sac, on the basis of his studies of the heart condition by the methods that have been mentioned. He does not hesitate to insert a needle and to withdraw the fluid.

The beat of the heart was first heard by an investigator who placed his ear directly on the chest. Then came the discovery of the method of listening to the heart through a solid rod. Later more delicate devices were discovered which had a bell for collecting the sound, a membrane for amplifying it, and rubber tubes to carry the tone of the sound to the ear. With the development of electric translation of sound and new means of amplification, it now becomes possible to step up the sound so that tiny variations may be determined.

Through examination after death from a known disease, it is possible to correlate all the observations made before death with the actual appearance of the heart and to learn the significance of each of the procedures used in diagnosis. Of special interest to the patient as well as to the doctor is the question of how capable the heart may be of carrying on its work. To the specialist in heart disease a dozen or more functional tests are known whereby he can measure the capacity of the organ to withstand unusual effort.

Heart disease is, to-day, the most frequent cause of death, its rate being at least twice that of the next nearest cause. Hence, the number of men giving special attention to diseases of the heart is increasing by the old law of supply and demand; though there are as yet few, if any, physicians who devote themselves exclusively to this branch. Diseases of the heart is a special department of internal medicine.

It is known to the specialist in heart disease that this organ may become inefficient and degenerate for many reasons. For instance, the acute infectious diseases may leave a weakened heart in their wake; and infections of the throat and of the teeth may be transferred to the valves

[201]

of the heart and to the lining of the heart with disastrous results. Venereal disorders sometimes produce strange changes in tissues of the heart. As man grows older, the strain on this muscle becomes greater and greater. It attempts to compensate for the strain by enlarging and by stretching, and its function is modified. There are many unsolved riddles yet to be answered before diseases of the heart will be brought fully under medical control.

No one knows exactly why the germs from the throat or teeth tend to localize in the heart in certain cases and to produce distinctive changes. This has to do, perhaps, with the constitution or hereditary stock of the individual. It may be concerned with the fact that the patient lives in a damp basement or near a river. It may be due to the fact that he is undernourished and has not had sufficient sunlight. It may be a certain form of a general group of germs that does this deadly damage rather than all of the germs of that group. It may be a combination of any such circumstances. Only when all of the facts are known and when the non-essential information is discarded will it be possible to prevent this type of heart disease and to supply proper treatment early enough to secure control. It is, of course, exceedingly important that patients come to the diagnostician soon enough to permit him to give proper advice. If the patient comes too late, such significant changes have already taken place that it is impossible to be of any real service. Education of the public is increasingly necessary to bring about frequent and careful examination of the condition of the heart.

In the past, patients with heart disease merely took to their beds and waited for the inevitable death, or continued their occupations until some too severe strain brought fatality. Modern study attempts to establish the exact capacity of the heart and to find for the individual an occupation suited to his condition.

With the heart, as with other portions of the body, it is extremely difficult to separate the effects that are purely mental from those that are physical. Associated with the shock of the World War were many cases of neurocirculatory asthenia, or weakness of the heart and of the nervous system, in which patients suffered from palpitation and pains and weakness without any detectable physical change being present. It is an extremely difficult diagnostic task to separate such cases from actual cases of heart disease.

In all of this work the newer devices that have been developed and the research that has come from the laboratories of physiology, pathology, and bacteriology throughout the world, have been of inestimable value. The endowments that continue to grow for furthering such investigation are going to make safer and happier men, women, and children in the future.

Diseases of the Lungs. The lungs, which are subject to many disorders, are now investigated not only by the old method, wherein the physician thumps and listens, but also by laboratory tests of the secretions from the lungs, by the use of the X-ray, by the insertion of a needle, and by the injection of substances which make the lungs particularly visible under the X-ray. Whereas the lung complaints most commonly considered in the past were bronchitis, pneumonia, and, most frequently of all, tuberculosis, to-day it is known that the lungs may be infected by many other bacterial organisms, that chronic diseases may produce permanent changes in the lungs which interfere with breathing, that the lungs may be subject to cavity, to consolidation, to the growth of tumors, to the inhaling of foreign bodies, and to many other conditions. When the diagnosis is properly made, treatment applied to the specific cause of the disturbance frequently results in the cure of cases which formerly passed inevitably to death.

[203]

The Stomach and the Intestines. The contents of the stomach may be pumped out and examined to determine whether or not food is being digested properly by the normal secretions. The shape of the stomach may be determined by injecting substances which make the stomach visible to the X-ray. A defect in the lining of the stomach shows on the X-ray and gives the physician an indication as to whether he must look for some abnormality in the shape of the stomach, for the presence of ulcers or of a cancer. Each of these conditions produces changes in the secretions of the organ itself. A certain amount of time is required for the stomach to mix the food thoroughly and to subject it to the various secretions, after which the food is passed on into the intestines. Any delay or undue haste in the emptying of the stomach manifests itself by symptoms which are exceedingly disagreeable.

By the use of the fluoroscope and by the use of serial X-ray pictures, it is possible to determine exactly how long the stomach requires for its work. By the use of the esophagoscope or gastroscope it is actually possible to look into the esophagus and the stomach. It is even possible, by means of a newly invented camera, to take pictures of the lining of the stomach without opening up the patient. In the vast majority of cases of indigestion, though, it is not necessary to make all these studies, for the simple reason that it is possible to tell from easily determined symptoms just about what is wrong and, by the giving of suitable advice concerning diet and digestion, to free the patient from his disagreeable symptoms.

On the other hand, in many cases called indigestion, it is difficult to make certain whether or not the patient is suffering from a disease of the stomach, a disease of the gall bladder, an inflammation of the appendix, or a nervous disorder without manifest changes in the stomach itself. There are people who find it impossible to eat or to swallow their food, owing to nervous manifestations which produce no definite visible change in the esophagus or in the stomach.

There are still many unexplained conditions involving the stomach which demand further research for their solution. There are instances in which the stomach fails to supply the proper secretions, and the reasons for this failure are unknown. Apparently the tissues of the stomach may be made to yield extracts essential to health and to life, such as the extract that controls pernicious anemia. Here is a new field for intensive research.

The procedures that apply to the study of the stomach apply also to the study of the intestines. Only recently has it become possible to pass a tube through the stomach and into the intestines in order to withdraw their secretions for investigation. Only recently have methods been devised for determining the rate at which food moves along the intestines, for studying the bacteria that live there normally and abnormally, and for observing the obstructions, paralyses, and many other disturbances which manifest themselves by serious symptoms.

There are forms of constipation which are due to interference with the digestion in the intestine, other forms which are due to disturbance of the motility in the intestine, and some forms which are merely the results of bad habits. There are situations in which it is exceedingly dangerous to eat food with too high an amount of roughage, and situations in which an increase in roughage provides the conditions necessary for a successful cure.

Within the past twenty-five years methods have been developed for changing the nature of the intestinal flora. Metchnikoff's original idea that the *Bacillus bulgaricus* was the normal inhabitant and that its presence was synonymous with long life has been changed to emphasis on another germ called the *B. acidophilus*. Attempts to implant the acidophilus organism in the intestine brought out the information that certain types of food were necessary to secure successful implants.

At the lower end of the intestine the development of varicose veins, commonly called hemorrhoids, represents an extremely disturbing condition. Fifty years ago only quacks attempted to treat this condition by injection methods. Their methods were found to be dangerous and not infrequently fatal. Yet continued study over a period of twenty-five years has developed several methods of injection, and particularly safe methods of prompt operation. The hemorrhoids are now removed under local anesthetics with prompt healing and with the loss of only a few days from work. They can, moreover, be removed by electrical desiccation as well as by injection methods.

Among the unsolved questions of medicine are the relationship of obstruction of the bowel to the development of serious symptoms of shock; the exact relationship of putrefaction in the bowel to the onset of various degenerative diseases; and, particularly, the control of infestation of the intestines by worms—conditions which are extremely common in tropical and oriental countries but which are now beginning to be found with increasing frequency in this country.

The Blood. Long ago it was recognized that the blood is one of the most important constituents of the human body, that it is the fluid which gives life to the tissues, that it takes from the tissues their waste products and conveys them to the organs of elimination, and that it is largely concerned with the ability of man to resist disease and to overcome damage to his body. Not, however, until the microscope was invented did scientific medicine begin to have any clear conception of the constituents of the blood. This fluid represents roughly from one-

[206]

thirteenth to one-twentieth of the body weight. It is far more than a fluid; it is a suspension containing many ingredients in the shape of formed and of dissolved elements which have an intimate bearing on life and health.

Metchnikoff conceived the idea that the white blood cells were the important element for attacking infectious disease. He gave to them the name "phagocyte," with the idea that they ate and digested cells. However, the white blood cells represent but one of the types of formed elements in the blood. There are in addition, the red blood cells and the platelets. The white blood cells number about 5,000 to 7,000 for an amount of blood about the size of a pinhead (1 cu. mm.). The red blood cells number from 4,500,000 to 5,500,000 for the same quantity. The blood platelets average from 150,000 to 250,000 for the same amount. Unquestionably all of these constituents have special functions, some of which have begun to be understood; but about many of their functions there is still a great deal of doubt.

The red blood cells seem to be charged primarily with the conveying of hemoglobin around the body. The hemoglobin contains iron and is concerned with the distribution of oxygen. If the hemoglobin is lessened in amount, there is less oxygen carried, and the blood becomes exceedingly pale. The normal amount of hemoglobin is listed at 100 per cent, but a person may have anywhere from 50 to 70 per cent and still feel fairly well. If the hemoglobin drops much below this amount, one is likely to indicate the deficiency by shortness of breath. The pallor associated with lack of hemoglobin is typical. There are many conditions in which the hemoglobin is altered or destroyed so that it cannot carry oxygen. For instance, in poisoning with carbon monoxid gas the hemoglobin is changed to a form which cannot carry oxygen, and the person becomes purple.

In the condition known as pernicious anemia there is a deficiency in the number and a change in the character of red blood cells. A lessened number of cells may be due to the fact that they are being destroyed too rapidly or to the fact that they are not being created sufficiently rapidly and in sufficient quantities. In pernicious anemia, their number decreases so rapidly and stays so low that it threatens life itself. The number of red blood cells may fall to 1,000,000 or 1,200,000, instead of the 4,000,000 to 5,000,000 that represent the normal condition.

One of the greatest finds of recent times was the discovery that the liver contains a substance which has the specific power of raising the number of red blood cells, apparently by stimulating the production of red cells in the places in the body where they are produced, notably in the bone marrow. Concentrated extracts have been developed which cause rapid formation of the cells; and a fatal termination to pernicious anemia, which was formerly inevitable, is no longer feared. Moreover, continued investigation now shows that there are substances also in the wall of the stomach and in the muscles which have this specific power.

Obviously, only the frontier of this unknown land has been invaded. Much more needs to be known as to the nature of the red blood cell, its method of formation, the materials from which it is formed, and the factors governing its formation, growth, and destruction. It probably has something to do with the ability of the body to resist the invasion of foreign substances. At all events, the red blood cells can be sensitized in such a way that they will clump together or will dissolve when certain substances of bacterial or other foreign origin are injected into the body; and obviously, if such reactions occur, the result to the patient's health and life is serious. The materials used in the treatment of pernicious anemia must also be investigated more thoroughly, and their functional ingredients determined. Already one can take a concentrated powder instead of eating large quantities of raw liver; and at least one liver extract is being prepared as a pure chemical.

The knowledge of the platelets in the blood is even more recent than that of the red cells. These platelets are seen only with great difficulty, and can be counted only by special technics which have been developed, none of which is of any great accuracy. The platelets are concerned with the ability of the blood to clot promptly when it escapes from a blood vessel. It is difficult to conceive of anything more important to a person's life than the ability to stop bleeding. Because of the importance of this reaction, nature has apparently developed a number of effective mechanisms. The blood contains a substance called fibrin, which is concerned in the formation of the clot. Unquestionably in association with the fluid constituents of the blood other materials may be involved. Most physiologists believe that the platelets have a prominent activity in this direction.

If the number of platelets is greatly reduced, the person bleeds easily under the skin. In some infectious diseases the number of platelets is promptly reduced, and one of the most significant manifestations of these diseases is the collection of large black and blue spots over the body.

The medical profession knows a good deal about the blood platelets, but there is far more unknown than known. It knows a great deal about the clotting of blood, but it needs to know much more. Tests have been devised which indicate that the blood clots normally in from one to three minutes after bleeding commences. The clotting can be hastened by squeezing the bleeding spot, by pressing on the blood vessel from which the blood comes, by applying hot applications, by applying substances which break down the cells, by rubbing, and by other manipulations with which almost everyone is familiar.

If there is a deficiency of calcium in the blood, this will delay the clotting. However, calcium itself is associated with many other conditions of health. Of these more will be said later.

The white blood cells are certainly as significant as the red blood cells for human health and life. When inflammation occurs anywhere in the human body, the blood brings enormous numbers of white blood cells to the spot. They come there and remove the degenerated or broken down material. They are of many varieties. In times of infection, for instance in appendicitis, the number of white blood cells will be raised promptly to 15,000 or even to 50,000 as compared with the 5,000 to 7,000 that is normal. One variety will be increased greatly in percentage as compared with the other varieties. This variety is called the polymorphonuclear white blood cell because it is a white cell with many nuclei. Whereas normally it composes from 55 to 65 per cent of all the white blood cells, the number may increase so rapidly in time of infection that this form will constitute 75 to 90 per cent. Another of the forms is mononuclear. In many diseases this form is greatly reduced, and in other diseases increased. A third form of white blood cell, called eosinophile because it stains readily with eosin, is increased in number in some diseases—unquestionably in infestation with the pork worm known as trichina.

The medical profession and workers in research laboratories throughout the world are spending vast amounts of time and energy to identify the various forms of white blood cells and to discover their exact significance in different kinds of diseases. Experts on the blood can tell even now whether a person is suffering from one type of infection or another, whether he has been poisoned by some chemical element, or whether his blood-forming organs are functioning properly. But though it is known that in the blood-forming and destroying system the bone marrow and the spleen are especially significant and that the liver and the gall bladder are seriously concerned, the whole mechanism is so complicated and intricate that it will require the best brains of thousands of scientists for many years before it is fully understood.

Blood Pressure. When a volume of fluid is forced through a tube, the pressure of the fluid in the tube varies according to the amount of pressure behind the fluid as it goes through the tube, and the size of tube through which the fluid flows. It varies also according to the ability of the tube to stretch. If the pressure behind the fluid is temporarily lessened, the pressure in the tube is lessened. If the tube is rigid, the amount of pressure will increase over that of a tube which can relax. These are some of the considerations which govern the blood pressure of man.

An increase in the amount of blood flowing through the blood vessels will raise the blood pressure; a decrease in the quantity will lower it. If the blood vessels become hardened by arteriosclerosis, the blood pressure increases. If the blood vessels are constricted by the use of certain drugs, or by certain reactions in the nervous system, the blood pressure increases.

The amount of blood pressure can be measured in numerous ways, including the method usually seen of putting a cuff around the arm, inflating it with air and reading the amount of pressure on a mercury column or on a spring device. Recently a machine has been developed that will record with a moving finger or moving screen the variations in the blood pressure over a certain period. By listening with his stethoscope while the blood pressure is being measured, the physician is able to determine two phases

[211]

of the blood pressure, while the heart is contracted and while the heart is relaxed.

At first everyone was concerned largely with the question of high blood pressure, but now it is recognized that either high or low blood pressure may be of diagnostic importance. Life insurance companies have given particular consideration to this question, and to-day the determination as to the life expectancy of an individual may rest largely on an exact determination of his blood pressure. The majority of medical directors of insurance companies are inclined to think that pressure somewhat below the average predisposes the individual to increased longevity. Pressure very much above the average is serious. These observations represent the fundamentals of our knowledge of the blood pressure, but do not begin to represent all of the knowledge that scientific medicine must have before it is able to render unequivocal judgments in this field.

Since the factors that control blood pressure are numerous and include not only purely physical factors but also mental and emotional conditions; since the part played by heredity is not certainly known; and since the influence of proper elimination from the kidneys, of the amount of fluid and of salt taken into the body, and of alcohol and tobacco, is not certainly established, there is a wide field of research on this frontier of medicine still available to the explorer.

It is believed that alcohol in moderation does not influence the blood pressure in some people. The withdrawal of excessive amounts of alcohol will lower the blood pressure in some people, but in others it apparently will not have this effect. Because of the food qualities of alcohol, it tends to favor putting on weight; and it has been established by insurance statistics that excess weight after middle life is likely to be associated with higher blood pressure.

[212]

It has likewise been established that a nervous state tends to raise the blood pressure, and it is known that excessive amounts of tea and coffee may make the individual more nervous than he would otherwise be. These practical aspects of the subject are such that every man may make, to some extent, his own diagnosis.

From time to time it has been recognized that various drugs will lower the blood pressure. This effect is brought about by action on some of the factors concerned in maintaining the blood pressure. Sometimes the decrease is brief and therefore of but little use in the control of a long-continued chronic condition. Other drugs have been used in which the reduction of pressure took place over longer periods of time. Then the question arises as to whether or not the increased blood pressure may not be a mechanism for maintaining the health rather than a serious factor of disease. Upon such problems as these research workers continue to spend their efforts in many hospitals.

The Liver and Gall Bladder. The liver is the largest organ in the human body. It weighs from three to four pounds. It is well supplied with blood, and it is concerned with many activities essential to life and to health. Under the surface of the liver is the gall bladder, a pear-shaped sac from three to four inches long. A tube comes from the liver and joins a similar tube coming from the gall bladder to make one common tube which empties into the intestines. The liver is the great chemical center of the body. When food is digested, it comes from the intestines to the liver, and the products of digestion are changed— built up or broken down—into other substances which are needed by the various tissues. Glucose, for example, is changed into a substance called glycogen, which is stored in the liver; later, the glycogen is reconverted into glucose and carried by the blood to the tissues as they need it.

[213]

The liver secretes the bile, which passes to the gall bladder and thence into the intestines. The liver helps to control the amount of sugar in the blood and to supply the tissues with glucose. It forms protein combinations needed in the body. It helps to rid the body of poisons, stores fat, and is concerned in the formation of chemical substances involved in the coagulation of the blood. No doubt there are many other functions of the liver that have not yet been suspected.

The bile is secreted by the liver continuously; and in animals that have a gall bladder, the bile is passed by it, from time to time, into the intestines. The amount of bile that will pass in twenty-four hours may vary from a pint to a quart and a half. Some of the factors which control the development and secretion of bile have been determined, but the complete mechanism is not yet fully understood. Moreover, the known functions of the bile are numerous, and many of them are still the subject of investigation. It is established, however, that the salts of the bile assist in the digestion and absorption of fat; that the bile helps to remove bacteria from the intestines through the slightly laxative effect which it has; and that in the bile are dissolved various toxic substances which are taken out of the blood by the liver.

If the liver of an animal is removed, the subject will remain normal from three to five hours. Shortly after, it becomes seriously ill, develops convulsions, the blood pressure drops tremendously, the heart beat increases, the breathing becomes disturbed, and the animal dies within two hours, unless given some treatment to prolong life. The same symptoms occur when there is a great fall in the amount of sugar in the blood, such as may occur, for instance, with a great overdose of insulin. It follows that deficient functioning of the liver may be associated with serious disease.

[214]

In their studies of the liver, scientists have developed methods for determining whether or not the liver is functioning properly. These involve injection into the body of a dye substance which the liver picks up and excretes. If the liver is functioning capably in the destruction of toxic substances, a larger amount of the dye will be eliminated than when it fails to function so efficiently. Since the number of toxic substances which may attack the human body is considerable, including germs, foreign proteins, and other poisons, the task of the liver in removing noxious compounds is one of the most important of the physiological functions.

So far as the gall bladder is concerned, extensive experimentation has already shown that it is quite possible for a person to get along fairly well without the organ. Many animals do not have a gall bladder, just as many animals do not have an appendix. The gall bladder appears simply to concentrate the bile by removing water from it.

The knowledge of this organ already possessed by scientific medicine has been of great value in saving human life, since a diseased gall bladder, or one that does not function properly, can be removed, and the person concerned saved from death by intoxication, or by infection of the distended organ, or indeed by rupture of the gall bladder itself. There are instances in which the gall bladder becomes filled with stones which may have, as their centres, germs that have gotten into the gall bladder and have been surrounded by concentrated salts. Such stones may block the passage from the gall bladder into the intestines, as a result of which the gall bladder becomes distended, and the person suffers great pain.

The intimate connection between the gall bladder, the liver, and the organs of digestion, brings about many complications; and a diagnosis as to whether or not a person is suffering from gall bladder inflammation, or

[215]

from an infected appendix, or from ulcers of the stomach, or from several other conditions, involves the use of many new and delicate tests, as well as a great deal of knowledge concerning the manifestations of the various diseases concerned.

In an earlier day the physician was compelled to base his examination of the condition of the gall bladder on what he could feel with his finger beneath the edge of the ribs. He still employs this measure; but, in addition, he is likely to make X-ray pictures to see if there are stones in the gall bladder; to inject substances which localize in the gall bladder and help to make it visible to the X-ray; to pass a tube through the stomach into the small intestines and to take out some of the bile for study; to examine the excretions of the body as to their content of bile and of bile salts; and to study the blood as to any changes which it may have undergone because of difficulties of excretion and secretion of bile.

Diseases of the liver involve gross changes in the organ due to inflammation, infection, poisoning by various metals which affect the liver particularly, hardening of the liver due to the effect of alcohol or changes brought about by food poisons, tumors of the liver, infection by tuberculosis or some of the venereal disorders. Since the liver is so intimately bound with the digestion and with the creation and maintenance of the blood, extensive studies of the blood and of the various secretions and excretions are necessary to make an exact diagnosis.

In the days when men used to drink tremendous quantities of alcoholic liquors day after day, cirrhosis or hardening of the liver with the development of hobnail surfaces was an extremely common complaint. As times and conditions have changed, this condition is seen more and more rarely.

On the other hand, modern industry has introduced the use of chemical substances of a poisonous nature

[216]

which frequently manifest their first serious effects on the human being by producing changes in the tissue of the liver and correspondingly prompt changes in the blood. Scientific investigators study these problems not only by chemical and physical studies of the human body, of the blood, and other fluids, but also by careful examination of the organs after death in cases in which fatalities have resulted from poisoning. In such instances the use of the microscope may show great changes in the blood cells that pass through the liver. The very tissue of the liver itself may have disappeared and have been substituted by fat. The metallic substances may be found in the cells of the organ itself. As a result of these studies, it becomes possible to plan methods of treatment that serve to relieve the symptoms, to save lives, and to extend the advance of knowledge still further into the realm of the unknown.

Metabolism. Before the coming of the modern era, the physician had no way of knowing exactly how well the vital organs of his patient actually were functioning. When it began to be realized that the body is a great physiochemical mechanism in which all sorts of reactions are going on, means were developed for measuring the rate of speed of these reactions and their functional efficiency. Everything that takes place in the human body is, to some extent, the result of a chemical change. Whenever any organ performs its duties, energy is used and is transformed into some equivalent of a different character. Energy comes from food. When the food is taken into the body, it is broken up into fundamental constituents which are taken up by various organs and transformed into compounds which the body can use. The sum of all these activities is the basal metabolism, which is influenced by many factors.

Formerly the value of all foods was expressed wholly as energy value, or caloric value. It was calculated that one gram of protein would provide about four calories, as would also one gram of carbohydrates. However, one gram of fat, which is more productive of energy than protein or carbohydrate, provides nine calories. It may therefore be seen how small a quantity of food is necessary to produce four calories and why it is that butter and fats help to put on weight so rapidly. Naturally, a person doing a small amount of work does not need to take in so many calories as one doing heavy labor. Whereas a clerical worker eats from 2,500 to 3,000 calories per day, a stevedore or woodchopper consumes from 4,500 to 6,000, and a lumberman may take as much as 8,000. In the same way, a seamstress takes from 2,000 to 2,300 calories, whereas a maid-of-all-work or a laundress consumes 2,800 to 3,500.

End Products of Digestion. Although much is known as to what happens to the foods when taken into the body, many additional investigations must be made before the complete story of the digestive chemistry is told.

Proteins break down into amino-acids and are circulated in the blood in this form. Some of these products are again put together to make body proteins, later to be broken down a second time under the influence of various conditions in health and disease. Some of the protein is changed into urea and can be found in the urine in this form. In certain diseases some of the products of digestion and body chemistry, which ordinarily are changed for various purposes within the body, are excreted unchanged; and these have been studied by the physiological chemist. By such studies it has been found, for example, that there are forms of sugar which may be excreted and which will give all of the ordinary tests for sugar, and yet prove not to be the particular sugar which is associated with diabetes.

A certain amount of various solutions is regularly found in the urine; but under certain conditions of disease, these solutions may be greatly increased in amount, or other substances may be excreted which are never found in the urine under normal conditions. Thus careful and complete chemical analysis of the urine leads to a better understanding of what is going on in the body. Quite logically, when any unusual substance is found in too great an amount, the tendency is for the physician to advise the patient to eat less of the substance which may give rise to the unusual material. By this means pathological conditions sometimes may be brought under control.

In an earlier day, the only examination made of the urine was to look at it in the light and to judge whatever might be judged with the unaided five senses. Gradually it has been realized that these excretions are an index of the body chemistry. The specimen may now be submitted to dozens, if not hundreds, of tests, the answers to each of which yield important knowledge. But here, again, only a beginning has been made; a vast amount of additional knowledge will be necessary before the whole truth is known.

Water. A good subject for a debating society with some knowledge of science would be the question as to whether water or oxygen is more important for the human body. Without oxygen one dies promptly, and without water, more slowly. Death from lack of oxygen is sudden and relatively painless; death from lack of water may be long and involve terrific upsets in the machinery of the human body.

Water carries materials into the body and out of it. As part of the blood it is concerned with interchange of materials within the body. Through its function of passing through membranes, it makes possible the continuous transfer of material from one cell to another and from one organ to another.

By the evaporation of water from the surface of the body the temperature of the human being is carefully regulated.

The German physiologist, Rubner, proved that a human being could lose 40 per cent of his body weight and still recover; but that serious trouble would follow a 10 per cent loss of water in the body, and that death after the loss of 20 per cent was certain. Obviously, the control of such an important substance in the human body, through the thousands of years in which the human being has developed, had to become an automatic procedure.

A person in good health is quite able to regulate the amount of water that he drinks. However, a man who is unconscious owing to brain hemorrhage, or who is paralyzed and unable to help himself, or who is ill in any manner which prevents him from satisfying his thirst, has to be looked after by some one else who must see to it that the supply of water is regularly maintained. Undoubtedly, in many instances of skull fracture or of brain hemorrhage or of shock, people have died because a sufficient amount of water was not available.

It has been estimated by several investigators that the amount of water put in represents about two quarts per day, usually taken in the form of drinking water; of water in coffee, milk, and soup; of water in solid foods; and of water developed by chemical changes within the body. It has been estimated that the amount of water passing out each day represents a relatively similar amount, short of about one-half pint. The water put out is in the usual excretions of the body, and particularly in water vaporized through the skin and through breathing.

The average man ought to drink at least eight glasses of water a day. He seldom realizes, however, that many of the foods that he takes must also supply additional water. Such apparently solid foods as steak, eggs, potatoes, oysters, tomatoes, asparagus, celery, and lettuce contain from 75 to 95 per cent of water.

The Kidney. The kidney is the organ most prominently concerned in the control of the water balance in the human body. It gets rid of the superfluous water when this is necessary, and sees to it that a proper amount of water remains in the blood when the supply is running low. If the kidney is diseased, the water balance may be upset and other diseases result. In various forms of diseases of the tissue generally, the method by which the body controls its water supply may be disturbed and the effects in the form of symptoms are prompt. If the body is deprived rapidly of water by fever, or if a person is unconscious and water is not put into the body, the blood will take up water from the tissues. The balance of water in the blood must be maintained, and even small changes in its content are accompanied by severe general reactions.

Of course, some water must be used in order to carry out poisonous waste products which may bring about death unless they are eliminated. Therefore, the kidneys may continue to get rid of some water until the whole body is pretty well dried out. It is an interesting commentary that prize fighters who attempt to make certain weights in order to fight in certain classes, subject themselves before the fight to this intensive drying-out process, and as a result sometimes enter the ring so completely shattered by the distortion of their water regulating system that defeat is inevitable.

In many diseases in which patients have been unable to take energizing foods so that their tissues have been exhausted in order to provide the body with these substances, the giving of small amounts of sugar with a large amount of water sometimes is followed by a prompt improvement.

Since the great importance of water in illness has been so definitely established, many methods have been worked out for making certain that any sick patient receives a

[221]

sufficient amount. It is possible to inject water into the body by the use of a syringe. Various fluids having the proper specific gravity and other desirable characteristics, have been developed which may even be injected directly into the blood. The fluid may also be put into the stomach with a tube, or into the intestines either from above or from below. If the fluid is put in gradually, it is taken up by the blood from the intestines and used by the body. If the person has a tendency to the development of acid because of his disease, bicarbonate of soda may be given with the fluid; but in many instances the mere giving of the water alone may suffice to clear away the difficulty.

Just as a person may be seriously sick from a lack of water in the body, a disturbance of his water metabolism may bring about a superfluity of fluid. The condition has been called dropsy by the public. It frequently represents a late stage of a condition that may be checked if treated early, but is handled with difficulty when treated later. Before there is recognizable swelling of the skin from excess water in the body, the patient may accumulate enough fluid to add several pounds to his weight. Thus the physician is likely to be suspicious of gains in weight which cannot be accounted for in any other way. If a person is accumulating extra fluids in his body, his skin usually has a shiny, transparent appearance, and, to use the jargon of the profession, "it pits on pressure." By this is meant the fact that when the finger is pushed against the skin, depression or pitting is produced which remains for some time after the pressure has been removed. The tissues feel distinctly doughy. If this edematous condition has persisted for many months, the skin is thickened and pits with difficulty.

Swelling of various portions of the body from accumulation of water may be due to a variety of causes. For instance, if the heart is unable to pump the blood to the ends of the blood vessels and back again to the heart, fluid begins to accumulate in the lower part of the body. If the person has a job like that of motorman or saleswoman, which compels him or her to be on the feet all day, the legs and feet will be found edematous. If the person is inactive and in bed, the thighs and the hips may be the parts most affected. Within the body itself the liver may be swollen, and fluid may accumulate in the abdominal cavity.

In the days before this situation was understood, it was customary to remove great quantities of fluid by tapping; and because the source of the trouble was not touched, the fluid that was removed was promptly replaced by a new accumulation. Operations have been developed for controlling this matter, but far more important is the treating of the diseased organ or tissue fundamentally responsible. It must be understood that the swelling due to an accumulation of water is a symptom and not a disease. It is important to know whether it is due to the heart or to the kidneys that are failing to function, and this can be determined only by the most efficient scientific investigation. When the fact is finally known, it is possible—by controlling the diet, by controlling the fluid intake, by the use of sweating particularly, and by various other methods—to bring about control.

Acids and Alkalis. Since the campaign of the California growers of oranges for the education of the public, the word "acidosis" has meant something to the majority of people. Somehow the word "acid" has developed a most unfavorable connotation in the human mind, while the word "alkali" is little understood by the average reader.

Acidosis is not a disease, but a disturbance of the relationship between acids and alkalis in the human body. Both types of substance occur in human tissues. If a sufficient amount of acid is retained in the body to increase the hydrogen-ion concentration of the body fluids beyond normal limits, the person has acidosis. The human body is equipped with a remarkably efficient mechanism for regulating the relationship between acids and alkalis. The mechanism includes, first of all, the lungs, which rid the body of large amounts of acid in the form of carbon dioxide. The second part of the mechanism is the kidneys, which dispose of acid by excreting it in the fluids which pass out of the body. The third part is the salt content of the blood and of the tissues, which can take up limited amounts of acid or alkali with a view to maintaining the reaction of the blood at a constant point. This salt content is called a buffer mixture, because it acts as a buffer between the upper and lower limitations of danger in relationship to the reaction of the blood.

Anyone with even a simple knowledge of machinery or mechanics will realize, therefore, that the human system is constructed with factors of safety against most ordinary disturbances. Conditions may arise, however, in which the buffers are used up; and reactions then occur which are unfavorable. In the presence of an insufficient amount of fluid or water, acidosis can develop, although such acidosis is very mild and is corrected by normal physiological reactions.

When acidosis becomes severe, nervousness, headache, irritability, nausea, weakness, and sleeplessness develop. The person seems short of breath and breathes with difficulty. At first he may be flushed and excited, but later, pale and exhausted. Sometimes there is a fruity odor to the breath, although this usually represents an advanced condition.

In some diseases, such as diabetes or Bright's disease, in very high fever, with profuse diarrhea, and with difficulty in the elimination of carbon dioxide in the lungs— such as occurs in pneumonia and heart disease—acidosis may be so severe as to represent one of the main factors in the cause of death.

In its advance into the unknown, scientific medicine has taken advantage of the work of the chemist for determining exactly the condition of the human body. The technical methods are extremely intricate and can be performed only by those who have had special training. They reveal definitely the exact extent of the acidosis from which the patient may be suffering and indicate also the nature of the treatment that must be used to overcome the condition.

Obviously, when acidosis occurs in diabetes or in Bright's disease, attention must be directed to improve the work of the organs that are weak, namely, the pancreas and the kidney. The work of the pancreas is improved by giving insulin and by changing the diet. The work of the kidney is improved by the use of sweat baths which eliminate materials through the skin and by a change in diet so as to throw less stress on the kidney.

In every one of the conditions in which acidosis develops, attention must be given to the water exchange, and it must be seen that the patient has a sufficient amount of water to take care of all of the factors that have been mentioned in the discussion of water.

There exists also the possibility of correcting the acid state to some extent by the use of foods that tend more toward the alkaline side. There are many foods which have value for this purpose. They include particularly peas, oranges, potatoes, peaches, cantaloupes, celery, carrots, beets, and lima beans. There are other foods which tend particularly toward the production of acid. Among them are bread, eggs, meats, oysters, oatmeal, and rice.

If severe acidosis is present, the person must be kept absolutely at rest, and care must be taken to see that he is kept warm.

[225]

The opposite condition to acidosis is alkalosis. This occurs rarely and usually because the person has been taking too much alkali or too much baking soda or bicarbonate of soda to overcome what he thinks is an acid condition. Many people who suffer occasionally from ulcers of the stomach or from hyperacidity, take immense quantities of soda for the correction of the condition. Obviously, alkalosis can easily be controlled by cutting down on the amount of alkalis being taken, both in the form of drugs and foods.

When alkalosis is due to excessive breathing, such as occasionally occurs in hysteria and other diseases, the patient may be given air to breathe which contains an abnormally large amount of carbon dioxide until the system is brought under control.

The Body Weight. It might seem, from all of the discussion and disturbance that has taken place in recent years on the subject of overweight, that all of the problems of overweight have been fully settled by the investigators of medical science. There are, however, still many unsolved problems which disturb the physiologist as well as the clinician.

It has been argued that overweight is merely the result of bad psychological bookkeeping; in other words, that overweight is practically always due to overeating and that it nearly always can be controlled by proper diet. The majority of medical opinion to-day is against that point of view.

Obesity, or overweight, is usually due to an accumulation of large amounts of fat distributed in the places where fat is usually distributed, but particularly in the abdominal wall, so that the obese person develops the appearance indicating that "coming events cast their shadows before." Women normally have a little more fat under the skin than do men.

As a rule, the average person maintains for approximately ten years a certain weight, which varies hardly a pound from any one year to another. Obviously, there is in the human body a regulating mechanism for balancing the intake and output so that the weight will remain fairly constant.

If a person over-exercises, he eats more; if he takes less exercise, he feels less need for food and eats less. This is largely automatic. As one advances toward middle age, he begins to add a little to the store of body fat, at thirty-five years weighing approximately ten pounds more than at twenty-five, and at fifty, ten or twenty pounds more than at thirty-five. This takes place so regularly that it is consented to be normal. It is probably due to lessened activity, a lowered activity particularly of the glands, and perhaps to a more quiet life in every possible way.

There are people, however, who gain much more weight than has been described as the normal gain. There are some families which tend to be fat. Just as there are animals thin and fat, large and small, so also there are human beings of various shapes and sizes, owing to the heredity of the family and to racial type. A German woman tends to be fat, and a Japanese woman to be thin.

In the disease condition called obesity, the human being stores up surplus material taken in the form of food instead of eliminating it so as to maintain the body weight at normal for the particular age and height. People who eat too much, exercise too little, and who eat particularly large amounts of rich foods are bound to put on weight. Children in such families acquire the bad habits from their parents, and the tendency to overweight becomes established. People in desk jobs tend to overweight. Owing to constant pressure on the regulating mechanism, it may break down after a certain amount of time and thus establish a vicious circle. There is a considerable number of people whose glands are disordered in some fashion as to interfere greatly with their regulation of body weight.

Just as there are thin people who eat tremendous quantities of food and still remain thin, so also there are fat people who eat very small quantities of food and continue to put on weight. This form of obesity is not understood. There is some abnormality in the mechanism of the person's body which forces him to store up fat. Some Chicago investigators have observed that the metabolic rates in these people were constant. They have also shown that the obese person has a completely disordered mechanism so far as concerns the handling of various types of food.

One of the first attempts made by a scientific investigator when confronted with the care of a case of obesity is to find out exactly the type of obesity from which the person suffers. It is necessary for him to study the family history and to know whether or not other people in the family suffer in the same way. He has to look into the family habits, particularly the diet. He has to make tests of the rate at which the body consumes its food and digests it. He tests the activity of the various glands, and he attempts to estimate for the particular person concerned what might be called the ideal weight for health.

The best authorities are convinced that one should not lose more than two pounds a week, the average safe loss being from three to six pounds a month. By the safe methods one who is greatly overweight can take off from twenty to twenty-five pounds over a period of four or five months, and should then maintain this weight for three or four months before any further reduction is attempted. Intensive reduction of weight should be undertaken only when the person is able to remain in bed constantly during the period of intensive reduction. Certainly nothing intensive should be attempted without having the constant attention of a physician who will watch for the presence of any dangerous reactions. He will examine the excretions for the presence of sugar or albumin; he will study regularly the ability of the heart to carry on its work and in this way make the process safe.

It is now generally known that the taking of thyroid extract speeds up the chemical interchange that goes on in the body. However, the taking of thyroid extract is not without danger. Some of the patented products sold in drug stores for the reduction of weight contain liberal quantities of thyroid extracts, and serious results have been reported from the taking of such preparations. Never should thyroid extract be taken without careful super-vision by some one trained in observing the changes that go on in the human body.

As can be seen from these considerations, the problems connected with weight are not yet fully understood. Because of the nation-wide attempt at reduction which has been made recently, more attention has been given to such problems during the past five years than at any previous period. Thus much information has been acquired, and a beginning has been made in establishing the ideal weight for health and in developing methods for securing and maintaining that weight. Perhaps the problem will never be solved until human beings have learned the import-ance of heredity and eugenics in the building of sound tissues. It has been possible in the breeding of animals to develop horses of both the racing and draft-horse type. With the same amount of consideration, similar results could be accomplished for the human.

Rheumatism and Arthritis. One of the unsolved problems of medicine to-day is the cause, the prevention, and the specific method of treatment of the rheumatic diseases. The term "rheumatism" covers a wide variety of disorders, varying from the rheumatic fever of childhood—which attacks the heart and which usually proceeds to permanent crippling or death—to inflammatory rheumatic conditions affecting the joints, general infections which affect the joints, gout, and similar diseases.

Rheumatic conditions are responsible to-day for a vast amount of disability, because a person with inflamed joints is usually not able to walk very well. There is no question that germs from the throat, the nose, or the teeth may be carried by the blood to the joints and there set up secondary infections. Obviously, it would be desirable to develop some method of injection or some drug that could be taken by mouth that would overcome such infection. In most instances, however, all science can do is to help to put the body in such shape that it will resist or overcome the disease, since there is as yet no established specific method of preventing or controlling these infections.

In addition to the rheumatic infections and inflammations that are definitely associated with the poison of germs, inflammations of the joints sometimes arise through what appear to be chemical reactions in the body, such as the reaction to the injection of some protein substance to which the human being may be sensitive. There are certain conditions in which a joint, particularly the knee joint, suddenly fills with water which, after an interval, disappears; then, perhaps after two or three weeks, the joint will suddenly fill with water again and, after three or four days, once more recover. The cause of this condition, which is called intermittent hydrarthrosis, is not known, and the current method of treatment is to endeavor to produce some reaction in the body which will change the general condition.

Not only are investigators convinced that inflammations of the joints are due to the actual presence of the germs in the joint, but many of them are inclined to believe that the poisons developed by the germs localize in the joints and set up inflammation. Usually, in such conditions, the whole body is found to be in a debilitated condition. The blood is not in a normal state, and the reactions of the body to infection are not up to par.

Perhaps the greatest discovery of medical science in relationship to the control of such conditions has been the usefulness of physical methods of treatment. The application of heat—either through hot water, hot air, electric heat, or other methods—is usually beneficial. A competent expert can secure for the patient, through the proper use of drugs and physical methods, a considerable amount of comfort. Sometimes, by the injection of foreign protein substances, he is able to arouse the system to such an extent that the body will shake off the disease.

Infectious Diseases. Again and again it has been pointed out that the greatest step ever taken by medical science followed the discovery by the chemist Pasteur of the germ causation of disease. When the cause of a condition is known, it is possible to prevent it more certainly, to treat it more accurately, and indeed to eliminate it entirely, if one can obtain public cooperation. When the mode of transmission is quite certainly determined, the task becomes even less difficult. Thus, typhoid fever, which was formerly one of the great scourges of mankind, is now quite generally under control, and from many communities has been almost completely eliminated. The death rate from typhoid fever has fallen from 35.9 per 100,000 population in 1900 to 7.8 in 1920, and to an even lower figure in 1930. The death rate in large cities of the United States is less than 3 per 100,000. Of typhoid fever the cause is now known. It is understood that the disease is transmitted by infected food and water and by carriers of typhoid germs. A medical diagnosis exists in the form of a test of the blood which indicates fairly certainly whether or not the individual is subject to the disease.

There is not as yet any specific method of treatment for typhoid fever, however, that overcomes the disease immediately. It has not been possible to prepare by inoculation of animals any serum or vaccine that will overcome the organisms in the body. True, vaccine will build resistance against the disease in a normal individual; and it is advised, particularly when one is going to travel abroad or in the country or in any place of uncertain water supply, that he be inoculated against typhoid fever. After the disease is once established in the body, however, the vaccine does not cure it any quicker than the patient can be cured by the usual methods of treatment. Neither is there for typhoid fever, as there is for malaria or for syphilis, a drug which has a specific effect on the germ. Research is tending toward the attempt to discover such drugs or specific biologic methods of treatment, but success is not yet in sight.

The Pyogenic Infections. Whenever germs get into the body and begin to release their poisonous products, the human being reacts with fever, chills, and with an increase in the number of white blood cells. He seems tired, he may vomit; and sometimes the germs localize and set up small abscesses. There are many types of such germs. One group is known as the pyogenic group, because pus develops when they attack human tissue. The chief members of this group are the staphylococcus and the streptococcus. Whenever a sufficient number of these organisms get into the blood, the person suffers with sepsis, which is commonly referred to by the press as blood poisoning, although the same term is also used for a venereal disorder.

Usually the infection with the streptococcus or staphylococcus begins at some single point on the skin or on the mucous membranes. It may begin because a roughened edge of a collar irritates the skin at the back of the neck, the usual procedure being first a pimple, then a boil, then a carbuncle, and finally, perhaps, "blood poison." It may begin by a mere scratch from a pin or needle, into which the germs enter. Not infrequently it begins through cutting a corn with a badly cleaned knife or razor blade. When the body fails to develop resistance and to throw off the germs, they multiply rapidly and the sickness is severe. The pulse becomes rapid, the face becomes pale; not infrequently great spots of hemorrhage appear under the skin. Death may occur in twenty-four hours or may be delayed several weeks.

Thus far the chief method known for controlling such conditions is to release as much pus as possible at the point from which the infection started. Unfortunately, in some conditions the infection is deep in the body and cannot be reached. In other conditions, it is a blood infection from the first. What is badly needed from the point of view of medical research is some chemical substance or some preparation that can be injected into the blood or underneath the skin or put into the body in some other way, which will overcome the infection directly and thus prevent the death of the patient. It is, of course, possible by modern methods of treatment to remove the localized spot of infection, or to use non-specific serums or vaccines in the hope of stimulating resistance or perhaps of striking accidentally the germ that is responsible. This procedure is not, however, in any sense of the word, certain or scientific. The main treatment of modern medicine is to give small amounts of water and nourishment, to see to it that all of the patient's organs operate at their best efficiency, to control the fever by baths, and to secure rest for the patient by the use of proper sedatives.

Calcium Metabolism. The human body is made of many ingredients which are of varying proportions. The chief ingredients are carbon, hydrogen, nitrogen, and oxygen. These can be found to the extent of 38 liters of water, 20

BIOLOGY IN HUMAN AFFAIRS

kilograms of carbon, nearly 4 liters of ammonia, 1.5 kilograms of calcium, 800 grams of phosphorus, 250 grams of salt, 100 grams each of fluorine, sulphur, and saltpeter, 50 grams of magnesium, and smaller amounts of iron, manganese, aluminum, copper, lead, iodine, bromine, and similar substances.

The bones of the human body derive their hardness from the fact that they are made largely of the salts of calcium. Calcium is commonly referred to as lime. The average man has about two kilograms of calcium in his body, most of it within the bones and the teeth. From day to day, by the processes of life, some calcium is lost from the body, and it must be made up by new calcium taken in. The new calcium can be had best in milk. Beside the carbon, hydrogen, oxygen, and nitrogen that are contained in milk in the form of protein, carbohydrate, and fat, this fluid also contains calcium, phosphorus, sodium, chlorine, and other elements.

Whereas most of the body's calcium is in the bones and teeth, small amounts occur in every tissue. About one-tenth of a gram of calcium is present in each liter of blood. If the normal amount of circulating calcium is reduced, the human being promptly has symptoms of a disturbing character. Among these symptoms are the convulsive nervous phenomena called tetany. Infants with convulsive disorders are sometimes promptly relieved through the addition of a proper amount of calcium to the diet. Some physicians believe that many cases of hay fever and asthma can be benefited by the taking of calcium. Then, too, the element enters into the reactions associated with clotting of blood and the prevention of hemorrhage.

Antiseptics and Dye Substances. The history of medicine reveals growths and trends in epochs divided by new discoveries which point the way to new methods of investigation. When Pasteur announced the germ causation of

[234]

disease, bacteriologists gradually defined the specific germ causes of diphtheria, tuberculosis, pneumonia, typhoid fever, whooping cough, scarlet fever, plague, leprosy, and many other infectious diseases. There remain, however, a large number of diseases, which are infectious in character, but for which specific causes have not yet been determined. Smallpox and measles are two of the common diseases of which the exact cause is not yet known. Influenza and the common cold continue to be the subjects of research. It has been argued that the causes of these diseases are organisms that are so small that they cannot be seen with the best microscope. These are the filterable viruses, so-called because they will pass through a porous clay filter. Much remains to be learned concerning them.

When Paul Ehrlich discovered salvarsan, stimulus was given to chemotherapeutic investigations, and innumerable studies have been made in an attempt to develop specific chemical substances that will attack certain germs and drive them from the body. We know that certain oils, such as the oil of chenopodium, are effective against the hookworm; that chaulmoogra oil is specific for leprosy; that salvarsan and its derivatives attack the spirochete of syphilis. It remains to find specific chemical substances that will attack the common pus-forming bacteria and overcome invasion of the blood.

When Almroth Wright wrote his epoch-making study on the use of vaccines, great impetus was given to the use of the killed bodies of bacteria to stimulate resistance to disease and to overcome infection. Now preventive vaccines are available for typhoid fever. Modifications of such vaccines are also applicable to other disorders. In general, however, the use of vaccines as specific remedies has been disappointing.

The isolation of diphtheria toxin and the development of antitoxin has established knowledge of diseases in which germs gain their effects by their toxins. Here also there is much to be learned concerning the nature of the germs responsible for such diseases and the nature of the remedies to be used against them.

Whenever new discoveries are made in any field of science, they are promptly applied to the possibilities for the alleviation and the cure of disease. Recent years have seen two special trends of utmost importance, one having to do with the determination of the deficiency diseases and their control by vitamins; the other having to do with the effects of great physical forces.

It is now known that such diseases as xerophthalmia, pellagra, polyneuritis or beri beri, scurvy, and rickets may be due to the deficiency of certain vitamins in the diet. They occur when these vitamins are absent, and the condition is cured when the necessary vitamin is supplied. Little is known as to the effects of relative deficiency of vitamins, but unquestionably this is associated with the possibility of lessened resistance to infection and minor complaints not yet perfectly recognized. It seems likely, furthermore, that there must be many more vitamins than those already discovered, just as there are many more unknown specific organ substances having to do with the growth and nature of the human body.

From the thyroid gland, the pancreas, the parathyroid gland, the adrenals, the testis, and the liver, extracts have been developed which control diseases formerly considered fatal. Pernicious anemia is overcome by liver extract; myxedema by thyroid extract; Addison's disease by an extract from the adrenal. Unquestionably, each one of the glands secretes several, if not many, hormones; and the isolation of all of them not only as extracts, but also as pure chemicals, represents the necessary research of the near future.

[236]

So far as physical forces are concerned, it is now known that the ultraviolet rays of the sun have the power to produce within the body the vitamin D that may also be derived from food or from irradiated ergosterol. It is known that the X-ray has the power of inhibiting the growth of tissues and indeed of disintegrating the living organism. A ray has been developed which is, in every sense of the word, a "death ray," with the power of killing at a distance. By the power of the X-rays and the rays derived from radium, severe changes are brought about in the human body; and it seems likely that wild cells with unrestrained growth, such as occur in cancer, may ultimately be made amenable to the effects of such radiation.

It is necessary to conceive of the human cells as being constantly swayed by many forces. They are animated with the life that comes to them through heredity, they are modified by the kind of nutriment that is brought to them by the blood, they are influenced by the action of minute living organisms that are parasitic, and their entire nature may be changed by the effects of light, heat, electricity, and other forces. Wherever there are so many multiples involved, the possibilities reach to infinity.

Surgery. Sir Berkeley Moynihan emphasized recently his belief that surgery had accomplished technically everything that it might ever accomplish for mankind. He failed, however, to realize the fact that the technic of the surgeon may be vastly improved by the invention of new devices for severing, uniting, or otherwise controlling human tissue exactly as the senses of the internist have been aided by audiometers, microscopes, and stethoscopes. But even beyond the improvement of surgical technic by such devices as the electric knife, the chromium suture, and various types of dressing, is the development of a new type of surgery that is preventive and physiologic rather than mutilating and pathologic. Modern surgery seeks to do everything possible to maintain physiologically normal function. It can produce rest of various tissues by side-tracking their functions temporarily and restoring them when the pathologic condition has been brought under control. Thus, before operating on the large intestine, the surgeon brings the small intestines to the surface of the body and sidetracks the large intestine until the operation has been completed and the patient has recovered. Operations are now done in two and three and even four stages in order to bring them within the capacity of the patient to endure. Through the development of reconstructive and reparative surgery, limbs are removed with the definite idea of providing proper stumps for an artificial limb.

A new point of view has been developed regarding diseases of the skin formerly considered merely as blemishes on the surface. Diseases of the skin were treated with lotions and ointments. Now it is recognized that any manifestation on the skin is a reflection of some constitutional disturbance, and the basic cause is sought and treated, the external treatment being applied merely for the patient's comfort. In the treatment of skin diseases, all of the powerful forces that have been mentioned are utilized.

Conclusion

The field of medical research is unlimited. In the consideration of the application of modern medicine to the study of internal medicine, I have shown the trend of medical investigation. The measures mentioned are being applied to all the medical specialties. A detailed consideration would involve writing an encyclopedia.

It has been taken for granted that the man of to-day is the best type of man that could ever be produced. Actually, by the application of our modern knowledge of nutrition, it may be possible to make men who are bigger, more resistant to disease, and even better thinkers than man of to-day. Through the application of biochemistry, physics, physiological chemistry, and all of the many sciences on which modern medical science is based, diseases will be brought under control, and perhaps many conditions anticipated before they have reached serious proportions. The frontiers of medicine, however, since it is a science of living human beings, will never be wholly conquered, for they are constantly extended. Diseases and the living organisms associated with the cause and the transmission of disease constantly change. Medicine as a living vital science must move forward, must advance, if it is to keep abreast of the enemy.

Chapter VIII

THE OUTLOOK OF PUBLIC HEALTH WORK

by Hugh S. Cumming *and* Arthur M. Stimson

THE inference that disease is practically coeval with the existence of life seems justified by the testimony of prehistoric animal and human remains. This evidence supports the speculation that even the earliest forms of life must have been subject to competitive, toxic, and nutritional stresses which they were not always able to withstand. Surely, for practical purposes, we may assume that disease has always been among us human beings. It thus takes its place among natural phenomena like sunlight and gravitation.

For a long time man refused to regard disease as a natural phenomenon. It presented such an affront to his assumed dominance over nature that he could conceive of it only as a visitation from supernatural sources. Only with the development of natural science within the past few hundred years has it become possible definitely to classify disease among the phenomena which are subject to natural law. In ways similar to those by which man has learned to employ gravitation and sunlight to his advantage, he is learning how to avoid the ravages of disease. These methods consist essentially of first learning by research the laws which control the operation of these phenomena, and then, by suitable conduct in accord with these laws, of securing the desired ends.

History records attempts at very early periods to escape disease both by individual and collective effort. The latter

[240]

represents the beginnings of public health work. In the absence of more than the meagerest knowledge of the nature or laws of disease, these efforts were seldom very effectual, never perhaps economical, and at times possibly even deleterious. At the present time, in instances where our knowledge is scanty, our efforts must be characterized by similar shortcomings.

Health is life without disease. Recently propagandists have attempted to distinguish a condition of positive health in which the possessor is not only not sick, but is super-healthy. It is only a question of where we draw the line. A condition of maximum ability to perform useful functions combined with adaptation to and enjoyment of environment, is certainly one of health, and anything less than this is, in a sense, disease.

Public health work is collective effort to secure a condition of health for all citizens of the area in which it is undertaken. Judged by the progress which public health work has made since its modern inception, it is one of the most rapidly advancing and profitable of human activities. The very fact that it has been highly successful in certain directions has removed the evidence of its achievements from the observation of the average person. Those of our citizens who are without a historical background do not realize that but for the public health work which has been accomplished they might, at the present time, be going through epidemics of Asiatic cholera, bubonic plague, or yellow fever. These and similar conquests which are really revolutionary in their scope are what cheer the workers in this field to renewed effort, and justify the support which the public none too generously gives to the continuation and expansion of the work.

A distinction must be made between personal hygiene and public health work. There are many persons who appear to think that if they individually carry out an ideal set of rules, they will be able to avoid illness and become or remain healthy. Undoubtedly they would further assume that if all individuals did the same, disease would disappear. This of course is an incorrect generalization based upon a part truth. Hygienic living by individuals is highly desirable, and tends to reduce their liability to illness, and their menace to others. It is one of the objectives of public health work to secure the observance of personal health practices by individuals. On the other hand, there are numerous factors in the induction and transfer of disease which cannot be dealt with by individual effort, at least not practically or economically. The man who lives in a community where malaria is prevalent may observe hygienic rules, dose himself with quinine, and stay behind screens after dusk, and he may thus for a period, and at considerable sacrifice, avoid infection; but in the long run he is likely to overlook some loophole in the scheme and come down with chills and fever. On the other hand, by well-directed community effort it is often possible to eradicate the infection from the area, so that no person living there, even if his personal habits were careless, could acquire the infection. Against typhoid fever again, public health measures, such as the providing of safe water and milk supplies and sewage disposal, insure a cheaper, more certain, and more general protection than could ever be expected from the exercise of individual precautions by members of the population concerned. Again, in the case of some pestilences such as yellow fever and bubonic plague, the individual could hardly secure immunity except at the sacrifice of nearly all other activity, while public measures can be counted upon for speedy control of epidemics.

We have in the ancient Hebrew sanitary measures commonly attributed to Moses an excellent example of rules for personal health which, while limited in scope and often

[242]

redundant and confused with religious observances, undoubtedly made for better health of those observing them,
under the conditions then obtaining. Is this public health
or personal hygiene? In so far as it represents an attempt
by authority to influence the members of the population
to avoid disease by personal practices, it is public health
work, but the practices themselves will be found to be very
largely of the nature of personal health habits.

In contrast to this may be cited the huge engineering
undertakings of the ancient Romans, which had a sanitary
and salutary effect, whether we may be able to trace the
intent or not. The aqueducts, bringing presumably purer
water from relatively uncontaminated sources, and the
cloacae in turn removing the dangerous wastes, were
essentially public health works.

It would perhaps be difficult to cite any very significant
improvements in public health until within the past two
hundred years. Quarantine procedure, attributed to the
mediaeval Italians, must, in its original form, have been
an uneconomical and dubiously efficient measure. It
consisted in holding the personnel of suspectedly infected
vessels on board for forty days before permitting landing.
It was assumed that by this time all the susceptibles would
be dead or recovered.

The past two centuries, however, have been characterized by very rapid progress in both knowledge and practice.
Edward Jenner in 1798 laid the basis for the control of
smallpox, a disease which in his day disfigured or blinded
most of the population whom it did not kill. Specific
immunization, therefore, commonly regarded as among the
most recent weapons against disease, was really among the
first to be rationally and widely applied. The epidemio-
logical approach, also, whose modern revival with improved technic is often mistaken for its beginning, was
successful during the middle of the nineteenth century in

devising adequate, if cumbersome, methods of control of both typhoid fever and Asiatic cholera. Various poisons, apt to affect large groups—lead, arsenic, ergot—were also detected, and rational means for their control were devised. In nutritional disease nothing important was discovered until about the beginning of this century, with the exception of the significant fact that scurvy could be prevented and cured by feeding fruit juice or fresh vegetables. The mere classification of foods into their caloric content, useful enough in its way, ignored the much more significant content in vitamin, the discovery of which has revolutionized our notions of nutrition. The fact of infestation with certain animal parasites was recognized, but little was done to prevent it. On the social side, health organizations began to be established as it became apparent that something could be done in a governmental or community way to control at least some of the worst of the besetting diseases.

This very brief and cursory review brings us up to the period of the discovery of bacteria and their important rôle in the causation of sickness. Far more important for humanity than the demonstration of the bacterial cause of this or that particular disease, was the generalization by Pasteur that germs do not originate spontaneously, but are always the offspring of antecedent organisms of the same kind. It is on this generalization that one important branch of present-day public health work is based. Obviously, if germs could originate spontaneously in filth, water, or soil, their control would be virtually impossible. On the other hand, if they came only from ancestors in kind, these having their common habitats in the bodies of persons sick with particular and recognizable diseases, something might be done to prevent their spread to those who are well.

Public health work has had different beginnings in various environments. In England, for example, it is said to have originated in social reform. The abuses in working conditions and housing, arising from rapid and uncontrolled industrial changes, aroused the nation to the passage of legislation which, although aimed at the abuses, actually had the effect of public health acts. Typhus fever in New York, erroneously believed to be due to filth as such, led to the establishment of a health department and a campaign of sanitation. In Panama public health work was undertaken as an essential preliminary to building the canal. Elsewhere the work has been the outgrowth of hospital activities, of charitable efforts, of missionary endeavor, or of business enterprise. Sooner or later, however, it tends to become a function of government, the reasons for this being chiefly that in some of its aspects the exercise of police powers becomes necessary, and that the fixing of responsibility cannot be satisfactorily secured under other auspices.

In the United States the general order of establishment of governmental agencies for carrying out public health work has been: first, the large cities; second, the states; third, the small cities; fourth, the counties; lastly, the villages. The Federal Government under our Constitution can function only in international and interstate activities, but in these it finds a large and useful field. Public health duties have been assigned to it for only about half a century, but in one way or another it has contributed to the cause of public health for a much longer time.

The systems of organization which have been developed in different parts of the world for carrying out public health work still vary widely. In some countries the national government exerts a strong influence through its representatives, even down to the smallest communities. In others the central health office is little more than a supervising and recording device and a purveyor of safe water; the measures are carried out largely by private physicians

[245]

throughout the land. The measures themselves, however, are found to be very similar in countries of comparable geographic and cultural status. This latter fact is of course traceable to the general adoption of scientific methods in public health work. If based on tradition or mere speculation, public measures would of course vary extensively according to local influences.

In common with other human activities, public health work has undergone a progressively widening interpretation as to scope. Originally confined largely to dealing with only the most devastating pestilences, it became applied later to those diseases which we have always with us and, for that reason, are prone to disregard, but which in the aggregate are far more destructive than so-called epidemics. The unusual, the exotic, the sensational, create a public demand for action, which the insidious and commonplace do not. As long as disease was regarded as largely a matter of filth, public health endeavor was directed chiefly toward cleaning up visible accumulations of refuse and rubbish. To this sort of activity the term sanitation still clings. To a considerable portion of the public this is still the beginning and end of public health work. They call upon the health office to abate nuisances, but do not seek or follow its advice on matters of greater importance to their welfare.

When bacteriology as applied to medicine began to yield quantitative results, health students began to discriminate regarding filth. The invisible typhoid bacillus concealed in a sparklingly clear well water became vastly more important than the pile of ashes in the back yard. Distinctions were made as to whether nuisances were actual, potential, or negligible health menaces. Economies were thus effected in action, or former duties were turned over to other public agencies. The street cleaning department relieved the health department and permitted the latter to apply its scanty funds where they would do more good.

The revelations of bacteriology and parasitology opened up a vast field of research and endeavor, which has been intensively and very profitably tilled ever since. In fact, in spite of all that we know about the importance of other causes of disease—nutritional, toxic, psychic—it still seems fortunate that the infectious causes should have been the first to have been intensively studied by scientific methods. Most diseases present vicious circles, or a series of vicious circles. Infection leads to disease, disease to poverty and ignorance, possibly to crime; poverty and ignorance predispose to malnutrition, toxaemias, and psychic depression; and these, in turn, predispose to infection. The circle could be represented in various ways and could be amplified *ad infinitum*, but this illustrates what is meant. The most vulnerable point in this circle still appears to be the infection link. Its elimination has not been accomplished, but a very decided impression has been made in the short space of roughly half a century. A general death rate cut in half indicates that a powerful influence has been at work.

As has been hinted, the conception of the proper scope of health work has undergone a progressive broadening. Without relinquishing research or control measures in sanitation and communicable diseases, it has passed on to the consideration of other important aspects. The food of the masses has become recognized as a determining factor of no small importance to their health. The discovery of the vitamins gave as much impetus to activity along this line as the recognition of the rôle of bacteria did to that concerned with infection. For, although there might have been some degree of actual starvation among a submerged portion of the population, it was not the deficiency in calories which caused a rather widespread state of malnutri-

[247]

tion, especially among the lower age groups. Again, with the development of new industrial processes and the distribution all over the country of new articles and preparations, it was realized that a real health menace existed in the possible poisoning, both among the industrial groups and the general public, of the makers and users of some of these unfamiliar wares. Growing interest has also been manifested in mental disease as a cause of suffering and incapacitation, and the realization is at hand that something may and should be done to prevent or control this condition. Recently cancer has enlisted the consideration of public health officials as a condition deserving of, and to a degree amenable to, public health activities.

The student of public health is continually impressed with the immense significance of social and economic conditions to health, and health workers are increasingly employing an approach along lines suggested by this consideration.

With the foregoing summary by way of orientation, it is now necessary, in portraying the outlook of public health work, to examine in some detail various factors to which a brief allusion has been made.

Official public health work throughout the world is definitely committed to the acceptance and employment of the findings of science. This is not to say that there already exists a good scientific reason for every activity which is now carried out. There are still many gaps in knowledge which leave to the public health official the choice between doing nothing at all or doing what seems to him the most sensible thing, although he may not be able to quote solid scientific reasons for it. As soon as it becomes evident that these methods should be abandoned or replaced, or that they can be refined and improved, he is not, as a rule, slow to take appropriate action. For it must be remembered that as a public official he is always under the spurs of public opinion and contemporary criticism.

It is difficult to think of a branch of physical science which, if it does not have a bearing upon public health work to-day, may not have one to-morrow. Few would have thought, fifty years ago, that the work of an entomologist, patiently classifying mosquitoes as a labor of pure science, of an obscure monk observing the results of cross-fertilization of garden flowers, or of many a worker in chemical laboratories, engaged in replacing atoms in molecular chains, would now have commonplace applications in the daily work of health officials everywhere. No one can foresee what new and useful applications await future adoption.

Public health deals primarily with man as *Homo sapiens*. The diversity of belief and viewpoint which at least the American community presents from the religious, philosophical, traditional, and ethical standpoints makes it virtually impossible for the health official to regard his human population as anything more than a collection of more or less *intelligent animals*. If he were inclined to inject into practice any purely moral considerations of his own, he would soon be attacked from all sides. For example, he had better regard the moral and religious issues of birth control and of venereal disease as none of his business—that is, none of his public business—and stick to what scientific facts are available when it comes to practical application.

Now *Homo sapiens* is a biological entity, and it follows naturally that the science of greatest import to public health work is biology. Taking everything into consideration, it is probably true that biology has thus far contributed more to human welfare through public health work than any other science has. We are aware that chemistry has made great claims for recognition in this direction,

[249]

and that we could not go far in our daily work without its contributions. We confidently believe, moreover, that chemistry has only begun its beneficent revelations; but up to date it is believed that the foregoing statement is true.

Taken in a broad sense, biology includes a large number of branches. The space limits of this article permit reference to only some of the more striking applications of biological sciences in public health.

First of all, as regards man himself it is self-evident that all we can learn regarding his anatomy, physiology, and psychology is bound to have applications in a program which concerns itself with his health. The first two have been upon a solid scientific basis for many years and are being built up as fast as, or faster than, practical application can be made of their findings. Psychology, on the other hand, can hardly be described as more than emerging from the realms of a purely speculative discipline, although with the increase of interest which is everywhere manifest and the tendency to supplant introspection with experiment and comparative considerations, the outlook is most hopeful. Public Health has much need of psychology. These sciences which concern themselves with the so-called normal serve as the point of departure for the study of abnormal or diseased conditions in which we have the subjects of pathological anatomy, -physiology, and -psychology, and their numerous specialties.

Still dealing primarily with man, and of immense importance to public health, are certain considerations of sociology. In fact, it is approximately true that if sociology were able to solve some of the problems of man's communal existence, public health would find its problems reduced by half. For example, if the economic status of a portion of the population could be improved by 100 per cent, which it certainly needs to be, public health could guarantee the virtual elimination from it of such diseases as malaria and pellagra within a year or two.

The science of epidemiology, properly so designated within the past few decades, deals also primarily with man, although secondarily it draws in a great number of related subjects. It is a branch of demography—the study of disease in populations. Epidemiology, which in its methods deals first with the individual, and then assembles and analyzes similar data concerning a large number of individuals, has proved fruitful in recent years not only in permitting the formulation of generalizations concerning diseases, but also in pointing to the most profitable line of attack, both in research by laboratory methods and in the application of control measures. For example, epidemiology shows pellagra to be a disease essentially of the poor, not transmissible, seasonally limited, and associated with dietaries of scanty variety. Laboratory studies directed along the lines so suggested prove that the disease is caused by the deficiency of a particular food principle in the diet, and that it may be prevented and cured by providing foods containing that principle. As another example, poliomyelitis (infantile paralysis) may be cited. Epidemiology shows that it spreads in populations after the manner of infections, disregards social and economic status, and is characterized during epidemics by the production of an enormously disproportionate number of light cases and healthy carriers. On the basis of this information public health must recognize the futility of attempts to prevent its spread by restraining the liberties of individuals, and must concentrate on the early recognition of cases and the institution of appropriate treatment, and on research to improve or perfect present methods of specific prevention and treatment.

An enormous proportion of human disease is due to the fact that we live in a world containing other forms of life

[251]

with which we are in constant competition. These forms are both animal and vegetable, and they affect health in a great variety of ways. Poisonous snakes and plants, for example, have in some times and places been serious health menaces, and a great deal of study has been necessary in order to develop protective measures. With the advance of civilization, however, the emphasis has changed from these external and obvious dangers to the more insidious and difficult problems of parasitism. Parasites are animals or plants, readily visible or microscopic in size, which live in the body at its expense. In many cases they exert little harmful effect beyond diverting a portion of the food supply from the tissues. In others they produce poisons which irritate, disarrange, or even destroy the tissues of the host. The bacteria, for example, which are microscopic plants, are capable, according to their species, of producing such specific diseases as cholera, plague, typhoid fever, dysentery, pneumonia, and a host of others.

There is hardly any fact which can be ascertained by scientific means concerning the myriad forms of parasitic plants and animals which may not find some application in medicine and public health. Physicians and health officials, recognizing the importance of such knowledge to their specialties, have therefore supplemented the studies of zoologists and botanists by researches of their own, and in many branches have well repaid, by their contributions, the great debt which they owed to the biological sciences. For example, they have virtually created the subject of medical bacteriology, have made important contributions to entomology, and may safely claim as their own the sciences of serology and immunology.

The dependence of medicine and public health upon the biological sciences is so great that it is futile to attempt a list of even the special subjects concerned. The case may be briefed by the statement that without the sciences of zoology and botany there would be no public health.

The contributions of chemistry also are innumerable, and they have recently been ably and amply presented to the public in many attractive forms. It is sufficient to allude to the indispensable part played by chemistry in physiology in leading to an understanding of bodily functions, both normal and diseased; to the applications of chemical dyes in the study of bacterial and other parasites; and to the preparation of valuable drugs which play a great part in health work.

The third great division of the physical sciences is physics; and here, too, the contributions have been indispensable. For what would public health work amount to if there were no microscopes or other optical electrical and mechanical apparatus devised by physicists? How helpless the health official would be without the physical knowledge which he now employs in many of his major health undertakings: water purification, waste disposal, illumination, heating, ventilation, the pasteurization of milk. Again, the physics of the radiant spectrum is already playing an important part in diagnosis and therapeutics— both of which have important health applications—and promises a brilliant future of usefulness. The science of physical chemistry, according to the prognostications of many students of public health, is destined to make some of the most fundamental of all contributions to both theory and practice. It must be remembered that with the development of modern conceptions of the atom, the claim of chemistry to finality regarding the composition of matter has been challenged, and in so far as this challenge may be sustained, the more fundamental and far-reaching discoveries will be expected from the province of physics. That medicine and public health will share in these discoveries cannot be doubted.

And finally, in discussing the scientific basis of public health work, it is necessary to remember the intimate relation existing between that activity and what are called the medical sciences. Several of these have been referred to in passing, and it only remains to discuss relationships. The division of all science into subjects is, of course, only a convenience. It is understood that all experimentally verifiable knowledge forms a consistent whole, and that its classification into different disciplines is chiefly an acknowledgment of the limitations of the individual mind. If there were such a thing as a composite human mind, there would be little need of such classifications, for it would be able to follow the electron into the atom, the atom into the molecule, the molecule into protoplasm, the latter into the millions of known forms of life with all the various ramifications of their physiological processes, and the relations of these forms to each other and to their environment, without being conscious of the need for a change of subject on the way. And thus the difference between medical science and practice and public health work is only one of viewpoint and practical convenience. As has been stated, in some countries the medical practitioner is essentially the effective public health agent. He not only treats his patients for the diseases for which they consult him, but also advises them in hygienic living, administers protective inoculations, supervises bedside precautions to prevent the spread of disease, maintains isolation of contagious cases, and is otherwise active in carrying out a number of health precautions. In the event of an epidemic he becomes an important factor in control measures.

In this country, however, the attitude of the practitioner, at least until recently, has been that his public responsibilities were largely ended when he had reported cases of communicable disease occurring in his practice to the official health authority. Hence a sharper distinction arises here between medical and public health activities than necessity demands. The doctor's field has been predominatingly treatment—the health worker's, prevention.

This is not the place to discuss the advantages of the two systems further than to remark that the first, using the practitioner as the health agent, appears to work well in the small, compact, homogeneous countries where it is found, while the second system accomplishes comparably good results under the conditions which prevail widely in this country.

Under our system it is possible for the practitioner to fulfill his obligations without any extensive knowledge of, or experience in, public health work. On the contrary, it is not possible for a responsible health officer to perform his official duties without a thorough grounding in medical science, or at least the assistance of those of his staff who have had such training.

In summing up this section it may be said that public health work, at least of an official character, is definitely committed to a policy of practice based upon information derived from the basic physical sciences: biology, chemistry, and physics. Without this information it would, in all probability, be somewhat worse than useless. While a part of the data comes directly from the academic sources of scientific information, a large bulk is derived from those adaptations which are known as the medical sciences, and still another portion is the result of researches carried out by health agencies themselves.

It is the intent here to deal with public health agencies only to the extent that is necessary in order to afford an understanding of the outlook of public health work. The observations will be confined to conditions in the United States.

First of all, the agencies may be separated into two classes, the official and the voluntary. Official health agencies are appointed and maintained under law for the benefit of all the citizens of the areas concerned, and their expenses are defrayed out of the taxes. Their employees are responsible to the people for proper performance of duties. Their field usually comprises the whole range of health problems arising in their jurisdictions. They have legal authority to use police powers in the enforcement of health laws. Voluntary health agencies may be subject to local laws regarding incorporation, but are not responsible to the general public. They may engage in general health activities, but are more commonly devoted to one activity or one group of activities, as, for example, the control of cancer or the improvement of health in children. They sometimes apply only to selected groups of the population, such as the students of a university or the employees of a factory. The funds for voluntary health work do not emanate from the tax receipts, but are secured by bequests, donations, various propaganda devices, or the solicited or required contributions of the benefited clientele. They have no legal power to enforce measures.

Both official and voluntary health activities have in the past been subjected to some influences which have impeded or, to an extent, frustrated their usefulness. In the former, political preferment has perhaps been the greatest stumbling block. In the latter, exploitation by plausible but incompetent or unscrupulous job-holders has played its part. Happily the past few decades have witnessed a great improvement in both of these respects, but the memories and vestiges of these drawbacks have left their impress upon the outlook of public health work. Their return must be guarded against if popular confidence is to be deserved or retained.

Owing very largely to the hindrances which have been alluded to, there has been in the past some distrust and friction between certain official and voluntary health organizations. This situation also has been very much improved, and more and more the two classes of organizations are learning to cooperate with each other amicably and effectually. There were at one time a large and unnecessary number of voluntary organizations devoted to health work. They did not get along well with the official agencies or among themselves. Within the past decade there has been a beneficial condensation of these agencies into a much smaller number of well-organized and, in general, well-financed associations having definite objectives and able personnel. Their power for good has undoubtedly been greatly increased thereby.

The need for voluntary health agencies arises chiefly from the fact that there are a number of things which should be done in health work but which, in the present state of popular opinion, are not a proper charge upon the taxpayer. Much good has also been done in the past by calling public attention to certain needful activities, demonstrating the possibility of useful work along these lines, and eventually securing their promotion as a part of the official health work. On the other hand, an efficient health official, with his broad view of all the health needs of his community, will often discover a hiatus in the scheme which he is unable to fill, with the funds or authority at his command, but which a voluntary organization could fill admirably. It is becoming recognized as the best official practice in such contingencies for the health official to seek the aid of the voluntary association.

The organization of official health work is readily understood when once the Constitution of the United States and that of the particular state concerned are considered. Legislative and police powers reside in the states. A minimum of authority has been delegated to the Federal Government by the states, which, so far as health matters are concerned, extends only to international and interstate matters. The states have also delegated powers to cities within their respective boundaries by charter. In the case of some large cities the terms of the charters are so liberal that the local health authorities have become practically autonomous, subject only to the provisions of state law. The conditions in the counties and villages are neither so uniform nor so well defined in many states. In a good many of them the county health organization is practically an extension of that of the state, under the supervision, if not the administration, of a deputy official. In others the delegated powers of the county government enable them to maintain a health unit under the commissioners.

Considering, from this point on, only the official health agencies, it is necessary first to inquire to what extent they have been established and are being maintained proportionately to the need for their services. On this point there would be unanimity of opinion among students and practitioners of public health. Provision is still lamentably inadequate in the United States. On the other hand, the trend of the past twenty years has been most encouraging. Within this period the remaining states and important cities have provided themselves with at least the framework of a health organization, which might be filled out in succeeding years. A beginning also has been made during this time in provision for rural health service. Since nearly half of the population still lives in the country, the neglect of the rural dweller has been a serious oversight in public health administration. There has been an increasing interest in this phase of health work which has already borne fruit in the establishment of county health organizations in many of the states. This, however, has been a slow process, so that only a fifth of the counties in the country are provided with official health service, except such as may be secured in emergency from the state organization. This means that a vast number of American citizens are born and live their entire lives without the benefits of health service. There is no one to suggest to the women that they can secure healthier babies by following some simple rules during their pregnancies, no one to insure competent attention during confinement. It is nobody's business to encourage breast feeding and, later on, suitable dietaries for the babies, or to attend to immunization against diphtheria and smallpox during the first year of life. No clinic exists where children may be taken for examinations and advice and for the correction of physical defects. Nobody urges the care of the teeth, and their neglect soon leads to serious impairments of digestion or to worse ailments. And so on up through the periods of infancy, the pre-school period, the school ages, adolescence, and into adult life, there is no person in these neglected areas whose duty it is to bring the lessons and the benefits of well-established health knowledge to the attention of the people, or to provide facilities for its application of which they may avail themselves. Of course, where the local doctors are not already overworked, where they are well informed, and are, in addition, altruistic even beyond the average for their profession, some of these necessary things may get done; but the point is that it is nobody's responsibility to do them. The results of this neglect are evident to anyone who wishes to investigate, even though he lacks a medical training. The proportion of crippled, blinded, deaf, and poorly nourished children is high. Much school attendance is lost on account of sickness. Since the adults are the products of similar conditions, men and women stoop-shouldered and old at thirty-five are commonly seen. Their teeth are reduced to a few loose snags. Typhoid fever is apt to be prevalent, and in some parts of the country hookworm disease and malaria take their toll both of life and of productivity. Tuberculosis finds good soil to work on, and parents, since no one has told them better, subject their children to an exposure which is certain to lead to their infection and subsequent suffering or premature death.

This gloomy picture may seem overdrawn to those city dwellers who visualize the country as a blessed spot where happiness and health flourish. The picture may be overdrawn for some rural areas still unprovided with local health service, but especially favored by climate, natural resources, and economic conditions which enable the inhabitants, by reason of their prosperity, to avoid some of the menaces of ill health through their own efforts and through the medical attention which they are able to secure. The picture is not beyond the facts, however, as regards a significant proportion of some thirty or forty million rural inhabitants of this country.

Returning now to the city health organizations and those of the states, it is to be noted that the actual financial provision for their activities is everywhere but a small fraction of that for other civic activities fundamentally no more important, such as school maintenance, fire protection, police, etc. Everywhere health protection, as reflected in city and state appropriations, appears to have been an afterthought. This condition is due to two considerations: first, that considering the returns, official health work is not nearly so expensive as the other forms of service mentioned; and, second, that health work actually is still somewhat of an afterthought on the part of the average appropriating body. This is apparently a matter of individual psychology reflecting itself in concerted thought. The average man who is or believes himself to be in good health finds it irksome to think about ill health or its prevention. Only when he actually becomes recognizably sick does he develop an interest in health, and then it is of course too late for prevention.

The maximum effect of health work is thus never realized, because nowhere is financial provision made for

[260]

maximum effort. Every health official can furnish, at a moment's notice, a list of additional health activities beyond those in which he is engaged, upon which he could guarantee a more than 100 per cent return in service, if only the funds were forthcoming for carrying them out.

This condition in counties, cities, and states extends also to the Federal Government. The provisions for health activities on the part of the Public Health Service have never approached those made for other useful but no more necessary activities of government.

This leads up to the pronouncement of an outlook which, it may be taken for granted, is shared by students and practitioners of public health work everywhere. *Health is the birthright of every citizen.* In so far as he has been deprived of health by the neglect of his community and his government to provide public health service (as distinguished from purely medical service) he has been cheated of his birthright. No one questions the right of every child to an education and to the protection of the law. Relatively few, however, have visualized the importance of starting out in life with healthy bodies and minds, and the responsibility of civilization in securing them for every citizen, in so far as may be possible.

It has already been stated that the conception of the scope of public health work has constantly been extended. At first, people were concerned chiefly with dangers which presumably threatened them at the borders of their domains—strange terrifying scourges which must be kept out at all hazards. We still quarantine against smallpox and typhus fever from Europe, and properly so, although we have much more indigenous smallpox among us than do many of the countries against which we quarantine, and in spite of the fact that we have an apparently all-American brand of typhus fever at home which nobody gets excited about for long. But interest in our home-made disease

[261]

products has now been established for many years, and it extends throughout almost innumerable ramifications.

Public health work is still in process of discovering its material. It is not so many years since it perceived that children were an important part of the public. Before that they seem to have been neither seen nor heard. Nowadays they loom large in the vision and the activities of the average health office. Then someone discovered the mothers of the children, and presently the industrial worker was unearthed and shown to be worthy of consideration from a health standpoint. And so, as regards the conception of what constitutes the public, a continual broadening and shifting of viewpoint has been necessary. Perhaps the most recent class to be recognized as presenting special health problems which will have to be dealt with is the touring public, that very considerable portion of the population which is here to-day and several hundred miles away to-morrow, exposed, on the way, to a variety of give-and-take episodes with roadside brooks, refreshment stands, and other hazards.

It has been pointed out also how the primary interest was fortunately with the infectious diseases—those which are due to parasitism with microorganisms—but that diseases due to improper diet, to poisonous influences in the environment, to the rush of modern life, to psychic disturbances, to social maladjustments, and to heredity, have forced themselves upon the attention of health officials and students, and have demanded action for their relief.

To meet all of these varied demands calls for a great number of separate but coordinated activities, and it may be profitable to attempt a brief review and classification of them. Roughly, then, the activities of public health work may be divided into (1) research, (2) control, (3) education, (4) legislation.

Research in the past forms the basis of present-day practice. Present-day research is laying the groundwork of future progress. It is, or should be, a part of the program of every public health organization. Unfortunately, because of the inadequate financial support of many such organizations, actual research comes in for what little attention it can get after the imperative work has been done. Some of the state and city health organizations, however, have made notable contributions, through research, to the armamentarium of the public health worker. The Federal Government also, through the Public Health Service which maintains a personnel of some three hundred research employees, has contributed much valuable data. In addition we must remember the notable achievements of other branches of the government, of the universities, and of voluntary organizations devoted to health research or related activities. Moreover, as has been pointed out, new facts in any branch of physical science may prove to have immediate practical bearing on health work, so that such work anywhere throughout the world must be counted as a potential asset in health research.

Public health research is divisible into two categories: first, the fundamental sort, which seeks to discover and formulate the natural laws which underlie a great many phenomena relating to health subjects; and second, the practical sort, which attempts to solve some immediate and pressing difficulty. Both kinds are necessary. Enormous advances along a wide front follow the first kind, of which no better example can be cited than those which resulted in Pasteur's generalization regarding microorganisms. Those in the second category may be compared to the capture of a trench or a redoubt in military operations, but they have the advantage of greater assurance of permanency and of being repeated by other workers at other places.

The separate activities under public health research work are so numerous and varied that it would be entirely beyond the scope of this article to attempt an enumeration. Possibly some reference to the classes of personnel employed on this work may give an idea of its range. The United States Public Health Service, for example, maintains scientific workers of the following kinds: medical officers specializing in a great variety of clinical, laboratory, and field investigations; sanitary engineers, equally diversified in interests; chemists—organic, analytical, physical, physiological, biological, industrial, etc.; biologists—including various specialists in microbiology, parasitology and helminthology, entomology, planktonology, and limnology; physicists specializing in various branches of radiology, illumination, electricity, mechanics, etc.; bacteriologists and pathologists; sociologists, psychologists, and psychiatrists; epidemiologists, statisticians, and mathematicians. These and other scientists, with their technicians and assistants, make up a body of some three hundred persons. The library personnel, also, must not be forgotten, since it is essential in public health work to keep up with the literature in one's specialty.

Control operations of public health are a function of its official organizations which occupy most of their time. The "control" refers to the diseases which should be checked or eradicated. Naturally, the first step is to know where and in what numbers the cases of various disorders are occurring. A bureau of vital statistics is required to collect, classify, and arrange this material. Reports of communicable diseases are secured from practicing physicians, and are supplemented from various sources, such as the records of school and industrial absenteeism and the reports of visiting nurses. Then there are the birth and death statistics to be recorded and analyzed. From this mass of material, suitably tabulated and graphed, the health official of a great city, for example, is enabled to watch the progress of disease and the effect of his efforts from season to season and from year to year. To meet the needs as they arise he has a number of bureaus or sections devoted to inspection and sanitation, the control of communicable disease, visiting nurse activities, hospitalization, the health of children and mothers and of industrial workers, the operation of water works, sewage disposal, milk control, and a host of other activities.

For carrying out these measures the health official must be provided with legal authority, and in many instances this extends to police authority over the activities and liberties of citizens. It is customary in these days, however, to exhaust the possibilities of persuasion before resorting to the exercise of police powers. Persons who are innocently or otherwise maintaining a menace to the public health are usually given a chance to mend their ways voluntarily before the compelling force of authority is exercised. The health official realizes full well that his duty is to all the citizens, and that a friend and a convert is always an asset. There is unfortunately a very fair proportion of persons at large who, according to the verdict of the majority of their fellow citizens, are wrong-headed about these matters, and so obstinately so that there is no recourse other than to compel them to do certain things in the public interest.

Education of the public has long been a slogan amongst health workers. It is possible that the slogan is less heartily voiced in recent years than it was formerly. On the other hand, it is believed that actual public education in health matters has improved in the meanwhile. The fact is that the slogan was in some instances merely a confession of weakness. When a lack of success was encountered in health work, there was a tendency to blame it on the ignorance of the public and to insist that the public be educated. In many instances, however, it was found that the knowledge

[265]

which it was so ardently desired to transmit to the public was quite unimpressive, threadbare, and none too solid.

Nevertheless a great deal of good work has been done, if not in "educating" the public, at least in bringing the aims, programs, and needs of health work before it for a sympathetic appraisal. Practically every known method of publicity has been used; the poster, the handbill, the lecture, movie and radio, leaflets, pamphlets and booklets without number. Health information has been included in the curriculum of the schools and colleges, even including, to some extent, the medical colleges.

The opinion may be expressed, however, that something far more radical must happen to American educational methods before the general public is ready to receive the information which the public health profession desires to transmit. It is difficult to visualize this as taking place immediately or within a generation or so, since what is needed is less the acquisition of some of the facts of health and disease as mere disconnected information, than an orientation toward environment which is in accord with the revelations of physical science. It requires no more than a casual familiarity with the product of our schools and colleges to assure one that such an orientation does not at present prevail throughout more than a small proportion of the graduates. However well grounded they may be in those branches which are mysteriously designated the humanities, they are, in general, extremely ignorant of humanity, its relations to other life, to its own racial history, and to the problems of successful living. Although possibly proficient in some major subject of science, it is seldom that they have acquired a well-rounded view of relationships. Since it is manifestly impossible to raise up a generation composed principally or largely of persons well grounded in all branches of science, it may well be wondered what is intended in this discussion. What is intended is the inculcation of a different way of looking at things, less geocentric and less egocentric, which is the refined product of scientific thinking, and is, after all, the recognized true meaning of education.

Among educational authorities there are many advanced minds who realize the shortcomings of present methods. They labor under great difficulties in trying to overcome the prejudices of self-perpetuating pedagogy, and the indifference of a body of school teachers of whom so large a proportion regard their jobs as mere stop-gaps rather than life professions. Under these circumstances it is not strange that the product of the schools is not prepared in its mental outlook to become interested in, or to have convictions concerning, a subject apparently so remote as public health.

Much could be accomplished if really adequate courses in physical sciences and public and personal health topics could be introduced into the *normal* schools. A teacher convinced by personal demonstration in the laboratory and on field excursions, would be a more inspiring instructor than one who mechanically follows a textbook in which he or she has no real interest.

Legislation is no function of health authorities *per se*, but they have a duty in bringing to the attention of legislators matters so important to the public welfare as those pertaining to the public health, and furthermore in advising wisely as to the ends to be striven for and the means of accomplishing them.

The Public Health Outlook. Students and practitioners of public health have different outlooks, just as specialists in any other line of human activity. It is too much to expect that all of them would endorse what is said here; and in attempting to present the subject to readers presumably relatively unfamiliar with it, the best one can do is to give what is believed to represent the general attitude of the profession.

It has been deemed necessary to run the risk of wearying the reader with what is, after all, a very brief survey of public health work before attempting to summarize its outlook.

First, the public health official feels the confidence which is born of very real accomplishment in the past. Most of the great pestilences which formerly menaced mankind have been swept away, so far as civilized countries are concerned. We shall have no more devastating epidemics of bubonic plague, Asiatic cholera, or yellow fever in the United States. These diseases have yielded to science as they never would have yielded to any other attack. Many of the minor epidemic diseases are definitely in process of disappearance. Typhoid fever, for example, has been reduced by scientific attack within some twenty years in the proportion of from about seven to one. The paradox remains with regard to a number of other diseases that the means of getting rid of them are known, if we only had the means. Their persistence, in other words, is a social and economic, rather than a scientific reproach. To this category, so far as this country is concerned, belongs malaria, stated on high authority to be the only disease capable of making a geographic area uninhabitable for the human race.

There remains, however, a long list of disease conditions against which little or no headway has been made. There is at present no answer to the problem of influenza, the last of the great scourges to menace civilized populations. Of the minor epidemics, poliomyelitis and meningitis may be mentioned as examples which are little amenable to public health control. Meningitis is an example of a disease whose microbial cause has been known for many years, and offers a refutation of the common dictum that once the cause is known, the cure is in sight. This is due, of course, to the fact that every phenomenon has multiple causes. The one essential causative agent in meningitis is the meningococcus, but the contributory causes of epidemics have not yet been envisaged. As the health official surveys his public, he finds only a small percentage who, at any given time, can be described as in perfect health. On the average, each individual spends a week or more out of the year completely incapacitated on account of acute illness, and, in addition, a large proportion drag themselves around for much of the time in a state of reduced efficiency and well-being. The field for work is thus practically illimitable.

The health man has now no hesitation in turning to science for the solution of his problems. If he knows his history, he sees centuries of complete failure to control any disease on a non-scientific basis followed by the spectacular disappearance of a number of the worst scourges after the introduction of scientific control. The conquest of the remainder is to him merely a question of more science. Possibly this will have to be directed along lines as yet little explored, and, likely enough, progress will be held up pending the discovery of new methods of attack; but the major conclusion is clear—disease is conquerable by scientific means.

For this reason research has the hearty support, in theory at least, of all health workers. Though many of them are submerged in the day's task so deeply that they can seldom come up to take breath and look around, at least on these occasions they realize acutely how needful is more knowledge and how scanty, all told, is the provision for securing it. This research must be conducted not only in the laboratory, but also at the bedside and in the field. It must embrace not only immediate pressing problems but must probe deeper into general laws of nature. It must enter and cultivate fields at first sight apparently remote from health subjects. Existing research institutions of the governments —municipal, state, and Federal—should be encouraged and supported, as should those of educational and voluntary health organizations.

The health official cannot fail to feel the need of a greater knowledge and understanding of the public whom he is striving to protect against disease and help toward a more healthful life. Perhaps this is especially difficult to come by in America, where populations are so often heterogeneous as regards race, social and economic status, tradition, and other pertinent factors. He should also be keen to make known to his public what he is doing and planning to do in their behalf. He will often find, however, that the greater part of the population is unprepared, by reason of its lack of training and experience, to comprehend just what he is driving at. He will look forward to the time when popular education has habituated a considerable portion of the people to a scientifically oriented manner of thought, so that what he has to propound will be readily and convincingly assimilated.

A health worker of long experience will remember almost with wonder the expansion in the scope of health work within the period of his service, and be prepared for a similar broadening of conception in the future. To this end he will appreciate the need for the perfection and expansion of organization in order to deal effectively with new problems. He will realize the need for increasing support of his efforts, both financial and moral, and will cultivate more assiduously than ever the opportunities for cooperation with his co-workers in the public health field, and with all legitimate agencies which may need or offer a helping hand.

And yet, he will sometimes look into the future with a hope mingled with misgivings. His hope is based upon the demonstrated efficacy of the method of attack upon disease to which he has definitely committed himself. His misgivings may arise chiefly from doubts as to the direction in which his public is going. If it be upward further into the light which reason and science are shedding upon human affairs, he feels that his profession can make relatively short work of the worst features of disease prevalence. If the race is preparing for a retreat back into the unprofitable realms of outworn tradition, bigotry, unreasoning emotionalism, and mysticism, from which it had begun to emerge, the health worker may well feel that his efforts are not worth the trouble which they undoubtedly cost him.

When one recalls the wonderful advances which have been made within our own generation, there is every reason for an optimistic outlook for the future. In my opinion the progress of public health along logical lines depends largely upon the intelligent assistance and guidance to the movement from the medical and allied professions.

Chapter IX

PHYSIOLOGY OF TO-DAY

by E. KENNERLY MARSHALL, JR.

PHYSIOLOGY, in brief, is the study of function as distinguished from structure. When the physiologist looks at a living organism, he asks, "What does it do and how does it do it? What are its various parts for and how do they perform their functions?" The knowledge which has accumulated in regard to function is of two kinds, descriptive and theoretical. The first consists in statements of facts, as, for example, that the heart beats, that stimulation of the vagus nerve causes a slowing of the heart beat, or that muscular exercise causes an increase in the breathing and heart rate. Such observations, if accurately made, are incontrovertible, as they are mere description. They may be simply qualitative or may attempt to describe the phenomena in quantitative terms. Much of the older physiology was necessarily of this type. But after a certain number of facts have been accumulated, it is essential for the progress of science that they be correlated and that explanations for the observed facts be sought. The more recent physiology has, in large part, been of this theoretical or explanatory nature.

There are, in general, two modes of approach in attempting to explain the function of any particular organ, to unravel the processes concerned in the functioning of that part. One of these involves studying its variations in function in the intact normal animal, while the other involves more or less interference with the normal animal by isolat-

[272]

ing, or partly isolating, the particular organ in question. The former, which has been called the analytic method of experimentation in contrast to the latter, or synthetic, method, has the great advantages that normality is maintained—but at the expense of a control of variables—and furthermore, that it can be applied to the human subject. The latter method, while sacrificing normality, admits of a much better control of the variable factors which enter into the experiment.

In the past, a great deal of knowledge has accumulated concerning the function and mode of operation of the separate organs of the animal organism; but, owing probably to the use of the synthetic method, it is difficult to bring all these observations and generalizations into harmony with the physiology of the organism as a whole, of the normal, healthy, everyday man. And it is becoming increasingly evident that the physiology of the organism as a whole is not simply the sum of the functions of the constituent organs or parts. There exists among all the parts a finely balanced regulation, which always operates in such a way that the individual organs or systems cannot work separately in the body, but must be correlated together or integrated into a unity; for the animal must function as a unit—an individual. This phase of our subject is so vital that it will be well to illustrate what is meant by an example.

Let us see how an animal reacts to a change in its environment which demands movement—the seeking of food, fighting, or flight from some serious danger. Let us take an extreme case, such as where maintenance of life demands the most violent muscular movement of which the organism is capable. In the change from rest to violent muscular exercise two things are necessary for the acting muscles: oxygen and a readily oxidizable foodstuff to supply energy for the movement. To supply these, many organs or systems

[273]

besides those which are directly involved (muscular and nervous) must operate in a different manner from what they do at rest. Some must carry out their function much more rapidly than before; others must almost cease to function for the time being. The rate and depth of breathing may be increased many fold (in man, from six to one hundred liters of air per minute); this supplies the needed oxygen to the blood. But if the blood moved around with no more rapidity than when the body is at rest, the increased supply of oxygen at the lungs would be useless. Hence, the heart pumps the blood much more rapidly and in greater quantity than before; in man, the amount of blood put out in one minute from one ventricle may increase from about four liters to twenty-five liters or more. This is accomplished by an increase in both the number of strokes per minute and the size of each stroke. But at rest, only about one-third, or less, of the total blood pumped by the heart goes to the muscles, although they comprise about one-half of the total weight or volume of the body; so a further adjustment is necessary. This consists of a widening of the blood vessels in the muscles and, in fact, of an opening up of many stagnant blood capillaries so that the number of capillary vessels through which blood is flowing may be increased forty to one hundred times. But the opening of these vessels alone would not supply enough blood carrying oxygen to the muscle; and, what is more important, it would involve a marked lowering of blood pressure which would secondarily decrease the blood and oxygen supply to the brain—the master tissue of the body—a thing which must not happen if the body is to remain active. So what happens, in order that the muscles and brain may receive an adequate blood supply, is that the supply to all the other organs is greatly depleted by a decrease in the caliber of the blood vessels, and we find, under these conditions, that the intestines get so

[274]

little blood and oxygen that they almost cease to function, while the same thing has been shown for the kidneys, which almost cease to secrete urine.

It has also been found quite recently that the spleen contracts during muscular exercise and shunts considerable concentrated blood, which has been stored here, into the general circulation. This helps in two ways, by increasing the amount of circulating blood and by allowing the blood to carry more oxygen.

Even now, if the exercise is maximal, the contracting muscles will not get enough oxygen, so the body goes into oxygen debt—borrows oxygen, so to speak, on further credit; that is, does work without oxygen—and the blood may get into such an acid condition as is seen in disease just before death. After a bout of violent exercise of only a few minutes' duration, it may take an hour or so to pay back the oxygen debt and restore things to normal. In this example we have followed only briefly the integration necessary to supply sufficient oxygen to the active muscles.

Closely bound up with this integration of the various organs and parts, so that the body can act as a harmonious unit, is the stability and constancy of the organism. Claude Bernard may well be said to have formulated this idea in his theory of the constancy of the composition of the internal environment of the organism, the blood plasma and lymph. It is found that physiological mechanisms exist which tend to maintain the constancy of the organism when its normal state is disturbed. There takes place an increase or decrease of some function in the nature of an adaptation which tends to restore things to normal. The study of such adaptations would seem to be one of the central problems of physiology. Lawrence Henderson has written: "The law of adaptation in organisms, founded upon the fact of survival, seems to be quite as well established as the second law of thermodynamics, and almost equally serviceable.''

With this introduction, then, to what physiology is, we may ask what are the present problems which occupy its disciples, and what are its trends. Many problems are old, as old as the science itself. Thus, in the field of the circulation of the blood, many important problems which are being actively attacked were already thoroughly grasped and formulated by William Harvey, the father of scientific physiology. What happens is that with each generation of workers, with new methods and new angles of attack devised, the old problems are reattacked and pushed somewhat further towards a solution. Other questions appear quite new, but in many cases this is due to the fact that it has only been with the marked advance of the sciences, like physics and chemistry, that these problems could be properly formulated. The increasing importance of biochemistry to the study of living processes is quite obvious to anyone who has even superficially scanned this field. Many problems connected with these processes can be formulated, attacked, and solved only by using the technique of chemistry.

The marked advance of biochemistry as a method of analyzing living processes may be granted as one of the present trends of the physiology of to-day. The intimate nature of muscular contraction is much better understood as the result of chemical methods of investigation which have involved a quantitative analysis of what happens to the various compounds present in muscle during and after contraction. The whole story of the regulation of the acidity of the blood and the close relationship of blood and breathing as carried out at the lungs, has depended almost entirely upon biochemical technique for its present better understanding. The study of the glands of internal secretion and the hormones which they elaborate, has been one of the most fascinating stories of the last quarter of a century and has depended, to a large extent, upon chemical methods for its advance.

The results from the application of physical methods to the problems of physiology, although not so spectacular and generally recognized as those of the biochemical school, have nevertheless been of immense theoretical importance, and promise to be even more valuable in the future. The advances in the mechanical aspects of muscular contraction, the investigation of the heat production of isolated muscles under varying conditions, and the application of newer physical methods to the study of the nervous impulse and the electrical change in nerve during activity may serve as a few examples of this type of approach to physiological problems.

A third method of attack, the method of Harvey, is as old as our science itself, and still must hold an important place in the immediate future. This is what we may call the purely physiological method. Careful anatomical considerations, experiments on living animals, deductions from these experiments, more experiments to test these, and so forth, characterize it. Within the past decade or so, our knowledge of the importance of the capillaries in the dynamics of the circulation, of the functions and interrelations of many parts of the brain and central nervous system, and of the relation of the emotions to bodily functions—to mention only a few examples—has been enormously amplified by the use of the purely physiological method.

With the increasing interest, which we have mentioned, in regard to the physiology of the animal as a whole as opposed to a study of its systems and organs separately, a beginning has been made in building up what we may call human physiology, where the experiments have been carried out on man himself. In the fields of metabolism

[277]

and energy exchange of the body, of circulation, of respiration, of the physiology of muscular exercise, of digestion and of excretion, considerable bodies of facts exist for the human subject, and certain generalizations have been attempted. This trend in physiology is of tremendous importance to medicine, but has been necessarily limited by the lack of methods suitable for use on the human being. As such methods are being constantly sought and found, this side of physiology will probably yield a rich harvest.

Lastly, in considering the general methods of attack upon their problems used by physiologists, one sees clearly, although very faintly, a reconciliation being brought about with zoology. An increasing interest in so-called general physiology—or the physiology common to all organisms as opposed to the special physiology of any particular species—the awakening of zoologists to the functional side of their science, and the use of comparative physiological data to solve functional problems in the higher animals, all tend to indicate that a real comparative physiology may soon be flourishing along with other developments.

In a more concrete way, we may ask just what are the problems of function and just how are they attacked and partially solved? No better way of answering these questions can be found than by giving in some detail a few examples of how physiologists go about their job. These examples have been chosen from fields of work with which the author feels some familiarity; equally satisfactory examples could have been taken from many other provinces of the science.

Let us take first as an illustration a problem in connection with the circulation of the blood, a problem of first importance in this realm and one which was clearly grasped by the discoverer of the circulation himself. Thus, over three hundred years ago, Harvey wrote: "Meanwhile I know and state to all that the blood is transmitted sometimes in a larger amount, other times in a smaller, and that the blood circulates sometimes rapidly, sometimes slowly, according to temperament, age, external, or internal causes, normal or abnormal factors, sleep, rest, food, exercise, mental condition and such like." Harvey was here referring, of course, to the quantity of blood which circulated around the body in a given time, which is the same quantity as that pumped by the heart in the same time. The problem of the cardiac output is here clearly stated, and a qualitative expression of the main factors which influence the output is given.

Since Harvey's time, attempts have been made in various ways by a long line of investigators to determine the cardiac output of man and the variations of this function under different conditions. It is obvious, of course, that some indirect method must be devised; the output of the human heart cannot be measured directly. At first, guesses were made as to the amount of blood pumped out by the heart of man based on a measurement of the total capacity of the heart as determined on the cadaver. But, when it was later found from experiments on animals that the heart does not completely empty itself at each beat, the value of such estimates became doubtful. After the development of methods for keeping the isolated heart alive by perfusion, accurate direct measurements of the output of such a heart could be obtained. Based on the relative size of man and of the animal from which the heart was taken, a calculation could be made for the human cardiac output. However, although this line of investigation has added enormously to our knowledge of the function of the heart, it tells us rather what the heart can do than what it actually does in a normal man or animal.

With the heart remaining *in situ* in the intact animal, two methods have been extensively used to estimate the volume of blood pumped out. In one of these an ingenious apparatus, called a stromuhr, which is inserted into the aorta—or large artery leaving the heart—measures the amount of fluid flowing through it. In the other method, the heart is completely enclosed in an apparatus called a cardiometer, which can be easily constructed from an ordinary tennis ball and a piece of rubber, and is arranged to measure the volume changes of the heart. The change from its maximal volume to the minimal represents the amount of blood discharged at each beat; and since, therefore, the number of beats per minute is known, the amount discharged per minute can be easily calculated. The use of both of these methods has led to valuable knowledge concerning the activity of the heart, but, since extensive operative procedures are necessary, the question has been raised as to how far the values can be applied to the heart in the normal animal.

Over half a century ago, the German physicist and physiologist, Fick, announced a principle upon which many recent methods have been based. He did not attempt to try out a method based upon his idea, probably owing to the lack of suitable analytical methods for gas analysis in blood. The Fick principle, as it has come to be called, consists in determining the total quantities of oxygen or carbon dioxide in the blood of the right and left sides of the heart, respectively, and simultaneously estimating the quantity of oxygen absorbed or carbon dioxide given off by the animal through the lungs. Now, if M.V. represents the volume of blood in liters per minute discharged by one side of the heart, A_o the quantity of oxygen and A_c the quantity of carbon dioxide in the arterial blood, V_o the amount of oxygen and V_c the amount of carbon dioxide in the mixed venous blood of the right heart, O the oxygen

[280]

absorbed and C the carbon dioxide given off by the lungs per minute, we have

$$\text{M.V.} = \frac{O}{A_o - V_o} = \frac{C}{V_c - A_c}$$

Since all of these quantities except the required M.V. can be experimentally determined, the problem is theoretically solved. Technical difficulties, however, in determining the necessary values in the above equation have prevented its wide application even to animals and have been only recently surmounted in the case of the human subject.

The Fick principle has been used on animals, where blood samples can be taken directly from the right and left sides of the heart. In the human subject, it is a comparatively easy matter to determine the quantity of oxygen absorbed at the lungs or the quantity of carbon dioxide given off; such a determination is made in determining the so-called metabolic rate or basal metabolism so much used now in the diagnosis and treatment of thyroid disease. The content of oxygen or carbon dioxide in the arterial blood (that of the left side of the heart) can also be easily obtained by an indirect method of calculation and can be checked directly by obtaining blood for analysis from an artery, a procedure quite feasible in the human subject. The difficulty comes, however, in estimating the gaseous contents of the mixed venous blood of the right side of the heart, and because of this fact the simplicity and directness of the method as used on animals are lost. Blood, of course, can be easily obtained from various veins of the body, as is done every day in hospitals in making various blood tests; but such blood is not the same in its oxygen and carbon dioxide contents as that of the right heart. It has taken a quarter of a century of intensive research to work out satisfactory devices for determining the gas tensions, and hence the gaseous contents, of the mixed venous blood indirectly. All of these schemes

[281]

depend on equilibrating a gas mixture in the lungs with the mixed venous blood, analyzing it, and then finding how much oxygen or carbon dioxide the venous blood of the subject will contain at these gas tensions.

Because of the difficulty described above in applying the Fick principle to man, other avenues of approach to the problem were sought. Another type of method, based on a determination of the rate of absorption through the lungs of some foreign gas, has been investigated. The idea underlying this type of method is quite simple in principle. If we take some physiologically inert foreign gas which will pass into the blood in simple physical solution, determine its solubility in blood, and then find out how much of this gas is absorbed by the blood passing through the lungs from the right to the left side of the heart in a given short interval of time, we can obviously calculate how much blood must have passed through the lungs in that time to dissolve the observed amount of gas. This amount of blood flowing through the lungs represents the output of either side of the heart. In practice, however, this method bristles with difficulties and questions as to its accuracy. Thus, we must be sure that complete equilibrium is attained between the gas mixture in the lungs and the flowing blood, that the time allowed for the experiment is sufficiently short to prevent any return of blood containing the foreign gas; that the heart output is not changed by the procedure used, and so forth. Only quite recently have methods based upon this general principle been put upon a firm basis as to their accuracy.

Since the above type of method has many advantages as to accuracy and simplicity over that based upon the Fick principle, we may describe briefly just how a determination of the cardiac output is carried out with it. Of the foreign gases used—namely, nitrous oxide, ethylene, and acetylene

—the latter has proven the most satisfactory. The subject, after a period of rest or exposure to some special condition under which the determination is to be made, has his oxygen consumption determined by one of the usual methods. He then rebreathes from a rubber bag a mixture of acetylene and air for a period sufficiently long to bring about mixture in the lung-bag system (fifteen to eighteen seconds) and a sample of gas is taken from the system. After five seconds more, another sample is taken. An analysis of these two samples, the barometric pressure, and the oxygen consumption give all the data necessary for calculating the output of the right side of the heart, and hence also of the left side. The technical details cannot be discussed here.

What results have been obtained by the use of these cardiac output methods on man? In the first place, it has been found that a man in the basal condition (eighteen hours after food and at complete rest) has a cardiac output of 2.2 liters for each square meter of body surface. This means that, depending on the size of the individual, his heart pumps each minute from three to six liters or from nearly a gallon to a gallon and a half of blood. Moreover, at each beat of the heart a little over two ounces of blood is discharged. The cardiac output of any individual under standard conditions is a very nearly constant quantity, varying only slightly from day to day. Many influences, however, modify the amount of blood being circulated around the body. Thus, the ingestion of food increases the cardiac output, a large meal having a greater and more lasting effect than a small one. The emotional state of the subject is also of importance, for emotional disturbances may raise the value above the basal normal. In muscular exercise, the output of the heart may be enormously increased, and, of course, this is necessary to supply the active muscles with oxygen.

[283]

Many other quantitative data of great interest have also been obtained by the use of these methods for determining the output of the human heart, but there is still need for more information concerning variations of the normal under different physiological conditions, not to mention the application of the methods in disease. One of the most surprising conclusions, however, from the results already obtained is that the heart rate is no indication of the amount of blood which the organ is discharging per minute. Thus, a fast pulse may be accompanied by a decreased, and a slow pulse by an increased, output per minute.

Concerning the regulation of the cardiac output of the human subject, much remains to be determined. We know with certainty that ingestion of food, changes of external temperature, psychic disturbances, lowered oxygen tension in the inspired air, and muscular exercise all change the cardiac output, but we are at present far from understanding the mechanism by which these changes are brought about. It is, of course, obvious that the output of the heart cannot exceed its input; hence, where the cardiac output is increased, the input must be increased. But whether this increased venous return to the heart is brought about primarily by changes in the cardiac mechanism, by alterations in the blood vessels, or by both, cannot be stated with certainty at present.

The above brief description of how the problem of the cardiac output of man has been attacked presents an instance of what may be called quantitative descriptive physiology; as yet the explanatory or theoretical side has just begun. Let us now take another example from the different field where the trend of recent work has been more on the explanatory than on the descriptive side. We say that we can explain a mechanism or a process in the animal organism when we can express it in terms of known physical or chemical laws. In many cases, this cannot be done, and we can then only express the process in terms of what may be called "physiological laws." A good example of these two types of explanation will be found in recent work on the physiology of kidney excretion.

To understand the problem, a little must be explained about the intimate structure of the kidney. This organ is made up of many identical small units—in the human about one million in each kidney—called "renal tubules." Each tubule commences in a peculiar structure formed by an invagination of a tuft of very small blood vessels into its expanded end. This structure is called the renal corpuscle or glomerulus. The remainder of the renal unit is composed of a long tube formed of epithelial cells, which, in the mammal, are found by microscopic examination to be of at least three distinct types. This tube is designated the tubule proper, and has a markedly different structure from the glomerulus. Ever since Bowman, nearly a century ago, discovered the true structure of the kidney unit, physiologists have interested themselves in two related problems concerning the production of urine; namely, the assigning of the proper functions to each distinct part of the renal unit and the "explaining" how each part carried out its function.

All theories which have been proposed to explain the secretion of urine by the kidneys have assigned different functions to the glomerulus and to the tubule. It has been generally agreed that the glomerulus eliminates most of the water or fluid part of the urine, but there has been, until recently, quite a controversy as to the nature of the process by which fluid is elaborated by this structure. One school has held that the process occurring at the glomerulus is a physical filtration or ultra-filtration, such as can readily be carried out in a dead collodion or parchment membrane—in other words, a process explicable by known laws of physics and chemistry. On the other hand, some have maintained that glomerular function cannot be explained as a filtration, cannot be expressed at present in terms of physical or chemical laws, or duplicated outside the body. Since the living cells around the glomerulus are, on this view, supposed to furnish the energy for fluid elaboration, the glomerulus is said to "secrete" fluid.

Now, how has evidence been obtained favoring one or the other of these theories of glomerular function? We can only review here briefly some of the more important results of recent research, which seem at present to "prove" the filtration theory. The structure of the glomerulus makes it ideally adapted to function as a filter, and this argument from structure has been frequently used. As the rate at which filtration will take place across a membrane will depend, other things being equal, on the pressure-head available, the relation of urine formation to blood pressure has been studied. Two careful series of recent experiments have shown that the rate of urine formation by the kidney—the rate of fluid excretion—varies directly with the blood pressure. Again, if glomerular function is filtration, the fluid formed at the glomerulus should contain all the constituents of the blood plasma in the same concentrations as they exist in blood, except the plasma proteins to which the filter is impermeable. Quite recently a group of workers in Philadelphia have succeeded in collecting fluid from the frog's kidney immediately after it has passed through the glomerulus—a structure in this animal less than one hundredth of an inch in size. The fraction of a drop of fluid obtained has been subjected to analysis, and found to contain certain constituents in the same concentration as they are present in blood plasma, a finding strongly supporting the filtration theory of glomerular function.

In regard to the function or functions of the tubules, opinions have also been divided. Some have held that the cells of the tubules were reabsorptive; that is, substances passed from the lumen of the tube back through the cells to the lymph and blood stream; others have postulated the reverse process, a passage of substances from the blood and lymph into the lumen of the tubule—a process which is designated as secretion; and still others have assumed the tubules to be both reabsorptive and secretory in function. It has now been generally accepted by workers in this field that tubular reabsorption takes place; in fact, tubular reabsorption of some substances at least follows as a necessary corollary of glomerular filtration.

In regard to secretion by the renal tubule, however, there has been a long and active controversy for many years as to whether or not it occurs at all. Recently, experiments on the kidneys of two marine fish—lowly animals and apparently evolved for no other purpose than for physiological experimentation—in which only tubules and no glomeruli occur, have proven definitely that the renal tubule can secrete. How much tubular secretion occurs in the mammal and man, as well as in the different classes of vertebrates, must be determined by further work. Such research is in active progress and leads inevitably into the field of comparative physiology. It is possible that this line of work will also tell us something about the functions of the microscopically distinct parts of a single renal tubule.

It may well be asked if the above controversy about the mechanism by which the kidney eliminates urine is not purely academic, having no practical importance for mankind. Such is most certainly not the case. The kidneys are the organs of the body upon which falls the greatest rôle in keeping the composition of the blood plasma—the internal environment of the organism—at that degree of constancy which is essential to health. Stoppage of kidney function inevitably causes death. For a complete understanding of the problem of the kidneys, such questions as have been outlined must be solved, and it is only when the true solution is reached that we shall be able to interpret kidney disease in terms of physiology, and enable physicians better to cope with it.

Finally, let us take a third example. In no field of physiology has our knowledge grown more, have the results been more spectacular, or have the practical rewards to mankind been greater than in that associated with the hormones formed by the endocrine or ductless glands. Since the development of the doctrine of the bacterial origin of disease, probably nothing has stirred up modern medicine more than the doctrine of internal secretion. We are told that we are what we are—bodily, mentally, sexually, emotionally—largely as a result of a balance or unbalance of chemical products from the glands of internal secretion. Let us, then, review briefly the physiological side of the modern work upon one of these interesting organs. As so much has been recently written about the hormones of the adrenals, thyroid, and pancreas, we will take one of the less known, but no less important glands—the parathyroids.

The parathyroid glands are quite minute organs, four in number, and anatomically closely associated with the thyroid gland. The first clue to their function was obtained, as has been the case with many of the endocrine organs, by studying the effects of complete removal of these organs from an animal. This operation results in the manifestation of a condition known as tetany (*Tetania parathyreopriva*); this is in no way synonymous with tetanus. It was often observed in man as the result of the removal of thyroid tumors in the days before the physiological significance of the parathyroids was recognized. The symptoms of removal of these glands vary somewhat in different animals. In the dog, which has served for most experimental work, there is a stiffness of the muscles, associated with fibrillary contractions or tremors, and later cramp-like and clonic contractions, with eventually convulsive fits. Since death always results after complete extirpation of all parathyroid tissue in untreated animals, it is obvious that these small structures are essential for life.

The function of these glands remained obscure for a long time. About twenty years ago, it was found that after a removal of the parathyroid glands in dogs, the calcium content of the blood was greatly reduced in amount. This naturally led to an attempt to restore the calcium of the body to normal artificially, by injection of a calcium salt; and it was found that this treatment would relieve the condition of tetany and restore the animal to normal for a short period of time. The whole train of symptoms of tetany was held to be due to a lowered calcium content of the blood, which is normally controlled by the parathyroid glands.

The above theory of the relationship of the parathyroids to calcium metabolism did not, however, go unchallenged. Another theory gained considerable vogue for a while; namely, that the parathyroids act in the body as detoxifying agents and that the symptoms after their removal are due to the accumulation of toxic products normally rendered innocuous by them. Indeed, certain substances, called guanidine and methyl guanidine, were stated to occur in the blood of parathyroidectomized animals in abnormally large amounts.

The proof that the parathyroid glands perform their function by the production of a hormone which is intimately related to the metabolism of calcium has come only in the last few years. This proof consists in the extraction of the hormone from the glands of cattle—in a

crude form, to be sure—the use of this hormone to bring
about complete and lasting relief from the symptoms of
removal of the glands, and the demonstration that the
activity of the hormone is proportional to its calcium-
mobilizing effect. After removal of these glands in dogs, all
abnormal symptoms can be suppressed and life prolonged
indefinitely by repeated injection of the extract containing
the hormone. The effect is as remarkable as that of insulin
in diabetes. The calcium content of the blood of a normal
unoperated dog is also increased by injection of the hormone, which is similar to the action of insulin in decreasing
the blood sugar of a normal non-diabetic animal. This
action on the calcium of the normal has been utilized for
the biological standardization of the parathyroid hormone.

Concerning the chemical nature of this parathyroid
hormone, very little is known as yet. No doubt, success
will come in the future in the isolation of the active
principle as a pure chemical individual, as has already
happened for insulin and the hormones of the adrenal
medulla and the thyroid and the ovary. This will be a
great step forward, and will undoubtedly aid in solving
the fascinating problem as to how this hormone acts to
regulate the calcium content of the blood. The final
step in our knowledge of the hormone will be a determination of its chemical constitution and a synthesis in the
laboratory. We cannot do more than mention here that
this parathyroid hormone has already been used successfully
in the treatment of various conditions of tetany in children,
due, presumably, to an underfunctioning of the parathyroid
glands. It is possible that the hormone may be of service
in many conditions where the metabolism of calcium is
disturbed.

In conclusion, it is important to recognize the tremendous
amount of laborious work which must be done before even
small minor points in regard to the functioning of the

animal organism can be settled. The public hears of only
the final result, and even then only if it happens to be
something of paramount importance for the welfare of
mankind. The slow, steady growth of our knowledge of a
subject, contributed throughout the ages by hundreds
of workers, many now entirely forgotten, is never properly
appreciated. In this connection Starling has written:
"Every discovery, however important and apparently
epoch-making, is but the natural and inevitable outcome
of a vast mass of work, involving many failures, by a
host of different observers, so that if it is not made by
Brown this year, it will fall into the lap of Jones, or of
Jones and Robinson simultaneously, next year or the year
after." To which may be added what Abel has said: "the
lap into which the ripe fruit falls generally has a very good
head atop of it."

Chapter X

ZOOLOGY AND HUMAN WELFARE

by HOWARD M. PARSHLEY

I

WHAT is human welfare?

Deep thinkers have answered this question in a great variety of ways, each trying to catch the elusive essence of the Good, and few succeeding in such wise as others can understand and agree with. In recent years, however, a tendency is observable among philosophers to accept what Bertrand Russell calls the Good Life as the practical expression of human welfare and as the essential element in any valid definition of the phrase. From concern with abstractions, such as other-worldly ideals, the *summum bonum*, and the greatest good of the greatest number, we are thus brought to a primary regard for individual happiness, for the successful life of the individual. This point of view, once agreed upon, helps us to define human welfare in simple, honest, and scientific fashion and, in consequence, to detect and modify the factors which affect it.

Let us agree that welfare requires conditions of life that are reasonably favorable with regard to health, wealth, social relations, and intellectual outlook. Exceptional individuals, to be sure, may fare best under certain handicaps or may require some peculiar favor from fate; but certainly the conditions we have mentioned are, in general, the requisites for living the good life.

What is zoology?

The human race is a zoological species; it is one among the almost one million kinds of animals that have been scientifically studied; it is bound up with zoology as with no other science. The objective study of animals and the body of knowledge and doctrine derived from such study must therefore have peculiar importance for mankind, and we are here interested to see some of the ways in which zoology bears upon the phases of existence which we have distinguished as the necessary elements in human welfare. Zoology contributes to human health because the principles of hygiene and normal physiology are zoological and because many diseases are caused or carried by animals; to wealth because many animals are economically useful or injurious; to social relations because human reproduction and heredity are fundamentally animal in nature; to intellectual outlook because zoological theory forms an important part in our modern views of psychology, morals, and progress.

II

The aim of hygiene is to maintain conditions favorable to the normal activity of the human machine. Apart from the fragmentary and often dubious lessons of ordinary experience, this aim is accomplished by relying upon the discoveries of physiology, a branch of zoology which has made most of its important gains through studying the functions of organs in animals. By using other mammals, such as the dog, cat, and guinea pig, medical men can extend very materially the field of experimentation, which is strictly limited (though not entirely closed) when they are dealing with human cases. And the zoological fact that these animals are in many respects similar to the human species makes the results of such experimentation immediately applicable, in many instances, to the study of human physiology, normal or pathological.

[293]

In such work it is often necessary to inoculate animals with the germs of human disease, or to practice vivisection upon them, as the only method of understanding what goes on in the human body; and it seems incredible that any one should wish to obstruct this very necessary and remarkably successful form of investigation. Yet it is true that the anti-vivisectionists, obsessed by a sentimentally inverted humanitarianism, have sometimes lost sight of human welfare in their concern for what they suppose to be the sufferings of laboratory animals. It is becoming generally known, however, that investigators customarily use anaesthetics and otherwise treat their subjects with due care; and so, with the spread of popular knowledge, the extreme opponents of vivisection now appear to be losing the influence that not long ago threatened to retard the progress of medical science.

The diseases that are specifically zoological in nature constitute a forbidding menace to human welfare, some of the more important of them being syphilis, yellow fever, African sleeping sickness, dysentery, malaria, and hookworm disease. Each of these is produced by a specific form of animal life (except dysentery, which is produced by several different organisms), living as a parasite in the human body where it multiplies and gives off poisonous matter. The various effects of these animals and their waste materials constitute the symptoms of the diseases for which they are responsible. Healthy individuals become infected by a transfer of the living parasites from a person afflicted with the disease, sometimes directly, by contact, as in syphilis, sometimes through a more or less complicated series of intermediate circumstances, as in malaria and hookworm disease. All of the diseases just mentioned are destructive to a degree which is not commonly realized. For example, Chandler[1] states that "of all

[1] CHANDLER, A. C., "Animal Parasites and Human Disease," 3d ed., p. 147, John Wiley & Sons, Inc., New York, 1926.

[294]

human diseases there is none which is of more importance in the world to-day than malaria . . . It has been estimated to be the direct or indirect cause of over one-half the entire mortality of the human race. Sir Ronald Ross says that in India alone it is officially estimated that malaria kills over one million persons a year." The frightful effects of syphilis are commonly known in some degree, though not fully realized by most people; and hookworm disease has social consequences of tremendous importance.

Parasitism, the living of one organism on or in another, at the latter's expense, is a widespread phenomenon, of which these disease-producing parasites are especially striking examples. They happen to have great medical and popular interest, but the researches carried out in discovering and identifying them, in learning their life histories, and in getting a biological basis for successful treatment, are essentially zoological in character. To illustrate this point, we may look briefly at two of the diseases mentioned above.

Malaria is a common disease throughout the warmer portions of the world and has been recognized from early times as being especially prevalent near swamps and ponds. In its several varieties it causes feverish conditions in its victims and leads to anemia, general weakness, susceptibility to other diseases, and frequently to coma and death from obstruction of the circulation in the brain. Sometimes infected individuals recover and become immune; but, in general, the result of the disease, when it is common in a population, is to lower the average vitality and thus to produce sociological degeneration. There is a strong probability that certain ancient civilizations, especially the Grecian, fell into decline because of this disease; and the backward condition of many affected districts, even at the present time, seems certainly to have the same explanation.

[295]

Quinine is one of the very few genuine, specific cures known to medicine; and since its discovery and general use, it has enabled those who can get it to live safely in malarious regions. But real control and eradication of the disease had to wait upon zoological research. Until 1880 everyone supposed that malaria was due to air which was, in some mysterious way, "bad"; but in that year the parasite was discovered; and in 1898 Ross, in India, proved that the parasite can complete its life history and be transmitted into the human host only through the agency of certain definite species of mosquitoes. Thus, the combined efforts of the microscopist, the medical man, and the entomologist were necessarily involved in understanding one of the worst enemies of human welfare; and now the elimination of mosquitoes not only from houses but even from entire regions, such as the Panama Canal Zone, is accomplished under the recommendations of scientists who are primarily interested in the zoology—the anatomy, habits, and classification—of mosquitoes and their natural enemies. Great public works, such as the building and maintenance of the Panama Canal, on the one hand, and the comfort and economic success of whole populations on the other, are thus dependent, in large degree, upon zoological research.

Hookworm disease, called malcoeur, tuntun, miner's itch, and chlorosis, in various parts of the world, was known in antiquity and now afflicts hundreds of millions of people throughout the warm regions of the globe. Its victims are weak, lazy, shiftless, despondent—a most immoral lot. We who are without parasitic worms gnawing at our vitals can cheerfully cast the first stone and preach of sins and virtues . . . or, if we are zoologists, we can study the life history of *Necator americanus*, and thus provide medical men with the knowledge that will enable them to make good people out of bad. Just this has been done; and some remarkable discoveries have been made, such, for instance, as the relation between shoes and shiftlessness. We commonly assume that shiftlessness leads to going barefoot; but in hookworm regions the opposite is more likely to be the case.

The victim of a hookworm infection becomes weak and pale; growth and development are retarded, so that a person of twenty may appear to be twelve; in girls the breasts fail to grow; the face is bloated, the eyes staring; a craving to eat dirt becomes strong; dizziness, headache, profound stupidity, illiteracy, and weak resistance to other diseases appear as common symptoms; and of course industrial ability and efficiency are low, and a retarded civilization marks the place where from 30 to 100 per cent of the people harbor the parasites. The worms are about one-half inch in length, and thousands of them may live in the intestine of a single patient. There they cause unhealing sores from which blood is constantly lost; and they give off poisonous secretions which seem to cause most of these distressing symptoms. The eggs of the worms are dropped on the ground in the faeces of the host, where the young worms develop and live in the surface soil for six or eight weeks. Then they die . . . unless a barefoot boy happens along. If he steps on a young worm, it bores immediately into the skin of the foot, travels through the blood vessels, heart, lungs, esophagus, and stomach, and finally reaches the intestine, where growth is completed and reproduction occurs. Before long, the boy loses his tan and other attributes of vitality, becomes shiftless, and, when his turn comes, fails to provide his children with shoes, the greatest blessing that could have been given him when young. Thus it is that the lack of footwear makes shiftless people. All of this was quite unavoidable until the zoology of the hookworm was known, whereupon methods of control were clearly seen. First, prevent infection of the soil by the use of sanitary latrines; second, avoid infection of the person by wearing shoes and otherwise eliminating contact with the soil. Moreover, several very efficient drugs were found to rid the sufferer of worms already at home within him. But scientific, governmental care and insistence are scarcely sufficient to work the needed reforms in dyed-in-the-wool hookworm localities, such is the sloth and ignorance and poverty of the population; and only long continued efforts can apply successfully the benefits that science affords.

In recent years zoological research has made rapid advances in the department of physiology, particularly in regard to nutrition and the endocrine system. Some of the early results of this work proved to be of immediate practical value and inspired sensational treatment in the newspapers and in books of ephemeral popularity. We can easily recall various food faddists, the vitamine advertisers, Dr. Steinach's fountain of youth, "Black Oxen," the monkey glands, and other such headliners of yesterday; but the follies of modern publicity need not blind us to the underlying scientific realities.

The first necessity of human health is food; and it is in the precise constitution of food materials that modern nutrition studies have made remarkable discoveries of great theoretical and practical value. It has been shown that ordinary foodstuffs, to be sufficient for bodily welfare, must contain certain substances, very powerful in effect but very small in quantity, in addition to the proteins, carbohydrates, and fats which make up more than 99 per cent of their bulk. These substances are the vitamines, and no amount of food material, however good it may appear, will keep a mouse or a human being in normal condition, if it lacks any of these peculiar elements.

The vitamines are mysterious. Their exact nature and chemical constitution are by no means understood, but

[298]

some of their properties have been detected by means of extremely ingenious experimentation on living animals. In many zoological laboratories investigators who specialize on the physiology of nutrition are actively engaged in this work, rearing white rats, birds, and other animals on diets made up with reference to vitamine content; and new information about nutrition in general and about the causes of such diseases as scurvy, beri beri, certain eye troubles, and of failures in development and growth is rapidly accumulating. All this is set forth in detail in a later chapter; but it is mentioned here because it is a branch of zoological research which has a very direct bearing on human welfare.

Perhaps no department of zoological investigation has afforded more striking results than the investigation of the endocrine system. This is a group of peculiar organs in the vertebrate body popularly referred to as "glands" and more exactly known as ductless glands, glands of internal secretion, or endocrines. They have been recognized, in part, from early times; but only recently has their real importance come to be fully appreciated. The study of these structures is now being actively pursued, and, indeed, it constitutes one of the most exciting and difficult frontiers of biology, as well as one of the most promising in practical value. The endocrines, such as the thyroid, the parathyroids, the adrenals, and the sex glands, exert a powerful influence on the organism as a whole, through subtle substances called hormones, which enter the blood directly from the glands and are carried by it to various parts of the body.

It has long been known that removal of the testes in early life results in profound changes in the appearance and in the psychology of the individual thus treated, producing what is called the eunuch in the human species and, in other vertebrates, such types as the gelding, ox, and capon.

[299]

In general the result of this operation is to prevent the normal development of the usual male characteristics and to produce an individual which is mild, strong, and heavy, but incapable of reproduction. Such traits are obviously of value in the Eastern harem and in the use of animals for heavy work and for food. And where male sopranos are required for sacred music in religious worship, the choirs have been maintained (until very recently, at least) by surgical dedication of boys with musical talent. These facts led a few curious persons to experiment with various animals and with various organs, until Bayliss and Starling showed how an extract of the duodenum injected into the blood stream causes the pancreas to secrete, owing to the effect of a hypothetical substance of very specific and powerful nature to which they gave the name of hormone. The idea thus established, many zoologists applied themselves to exact and well-directed experimentation, and in a few years the modern science of endocrinology grew up, with a new terminology, a new technique, and a new knowledge and control of physiological processes which had hitherto been mysterious.

The thyroid gland, for example, produces a hormone which tends to maintain normal chemical activity and hence normal growth of body and mind. If this organ is defective, the lack of the hormone produces sluggishness and abnormal development that may result in various diseases, such as myxoedema, ordinary goitre, and cretinism —all more or less fully remediable by supplying thyroid substance from animals. If the organ is over-developed, exophthalmic goitre, great increase in basal metabolism, over-activity of the heart, loss of fat, and other changes occur, which can be advanced or retarded by various experimental means. Much of all this was discovered by research on tadpoles and other lower animals; and gradually it is leading to the scientific treatment of human beings, as the technique develops through experimentation on the higher mammals.

A most remarkable series of discoveries has resulted from recent investigation of the gonads—the testes in the male and the ovaries in the female; and here we have most decidedly an advancing frontier of zoology, for the facts already found point toward hitherto unsuspected complexities and bear directly upon functions of the greatest practical interest. The sex hormones certainly control the development of the secondary sexual characteristics, such as bodily size, strength, and shape; the distribution and length of hair; the quality of the voice; and probably certain psychological traits. The differences which naturally distinguish the sexes and play such an important part in sex attraction, mating, and reproduction are absent or imperfectly developed if the sex hormone is not produced at the normal time. There is little doubt that some cases, at least, of perverted and otherwise abnormal sex behavior depend upon endocrine disturbances, and it is equally likely that further knowledge of these conditions will lead to practical methods of cure. But there is much more than this.

In the female the ovary produces several hormones. One, as we have seen, governs sexual development; another, produced by the corpus luteum (a group of cells formed on the surface of the ovary as each egg is extruded), controls menstruation and affects the condition of the breasts most strikingly when pregnancy occurs. If the egg fails to develop in the uterus, the corpus luteum soon degenerates, menstrual bleeding follows, and the mammary glands are little affected. But when the fertilized egg is implanted and forms an embryo, the corpus luteum persists and grows larger, probably preventing menstruation and giving out the hormone that stimulates growth and activity in the breasts. There are obscure relations here

[301]

between at least three parts of the ovary—the interstitial
tissue or framework, the follicles in which the eggs
develop, and the corpora lutea—not to mention the lining
of the uterus; and exact knowledge about these relations
has not been reached. It is in such matters that experimentation on higher animals must be relied upon to form a
background for the rare opportunities that permit direct
work on human beings in the course of surgical practice.
It has been found, in a few cases, that grafting a piece of
normal ovary so that it grows in the body of the patient
will cure abnormal and painful menstruation; but the
significance of such facts in detail can rarely be made out
unless the operation can be repeated with various modifications on animals. Zoologists are actively engaged in these
studies in certain laboratories where the necessary equipment is provided; and they have developed a new branch
of technique, biological surgery, which enables them to
transfer living tissues from one animal into another,
perform operations on embryo mammals, observe the
condition of the reproductive organs at various times and
after specific treatments, remove single glands without
otherwise disturbing the animal, and in general carry out
ingenious and dangerous manipulations with a percentage
of success far greater than has been possible in the past.

This has made possible certain investigations on the
control of reproductive activity in the female, which are
now being quietly but actively pursued and which have
already yielded results that will, in all probability, have
great significance for human welfare. By the use of the
female hormones mentioned above, it is already possible
to render certain female animals either fertile or sterile
at will for given periods; and the successful application
of such methods to the human species can hardly be considered more than a mere matter of technique. With the
hormone material concentrated or isolated and thus

readily available and the effects thoroughly understood, birth control will become not only sure and easy, but also biologically sound in the sense that ovulation and menstruation will be started and stopped by the natural agency— but at such times as seem desirable to the persons concerned. Contraception is already well-nigh universal in spite of the imperfection of the commonly known methods; and endocrine zoology promises to make it safe and certain. If and when this happens, the dwindling forces of opposition may as well abandon the struggle and contribute their enthusiasm and altruism to the important task of seeing that reproduction is reduced and regulated in such manner as to benefit all concerned.

The most recent accomplishment in endocrinology is the preparation by Swingle and Pfiffner of an active extract of the adrenal cortex, distinct and free from adrenalin. This extract will keep alive cats whose adrenal glands have been removed and which would die at once without the artificially injected hormone. At the Mayo Clinic this extract has been successfully used in advanced cases of Addison's disease, restoring patients from a state of complete collapse to activity and "the appearance of perfect health." Thus another fatal disease is brought under control, as was diabetes a few years ago.

What has been said about the new zoological technique applies also to cancer research. Abnormal growths can now be transplanted from one mouse to another and otherwise studied more closely than has been possible in the past. One striking discovery that has been made in this work shows that certain cancers and tumors, when transplanted, will grow only in animals that belong to certain hereditary strains. If the mouse has a different hereditary constitution, the transplanted tissue will simply disappear. It is easy to see that such facts have a direct bearing on the whole question of constitutional susceptibility to disease, as well as on the problem of the incidence of cancer; and with the advances which these studies are bound to make there is every reason to suppose that we shall soon be able to deal with pathological conditions that are incurable and fatal in our present state of knowledge. For this we shall have to thank not only medical and surgical science, but also the zoological research which is fundamental to the success of clinical practice.

III

We have agreed that wealth is the second factor in human welfare, not as measured in millions, perhaps, but at least as represented in sufficient property and income to maintain us in reasonable comfort and to provide us with the opportunities and things that seem reasonably desirable. Many animals are of interest to us from this point of view; they destroy things we want, or they furnish things we can use. Now, economic zoology is the study of such animals, and it is not purely a matter of economics proper, because the effort to foster beneficial species and to destroy those which are injurious involves researches essentially zoological. The habits, life histories, physiology, classification, and distribution of economically important species must be known before practical measures can be formulated.

All civilized governments recognize the value of economic zoology and maintain more or less elaborate departments for its pursuit. Politicians come and go, but the scientists (who often fall temporarily under their dubious direction) remain, slowly advancing their laboratory, museum, and field work and contributing, after their fashion, to whatever measure of prosperity the country may enjoy under varying administrations. It is well for the public to realize this essential continuity underlying political change, and it would be to the people's interest

[304]

if they gave more generous support to the government scientists and less superficial attention to administrative job-holders. It will astonish any able and successful business man to learn of the miserable salaries and appropriations under which many scientific departments must labor. Unless some popular branch of science is concerned, such as radio, legislative neglect and parsimony seem to be the rule. This situation could be changed, if sufficient pressure were brought to bear by interested and informed citizens. If such pressure should bring about increased appropriations for scientific work, we could be sure, at least, that the money would not be wasted in graft and display. For scientists are interested in science, and they are always seeking to improve their equipment and extend their researches.

One of the most important of the scientific departments of the United States Government is the Bureau of Entomology, which is devoted to the study of beneficial and injurious insects. Its practical work was begun by C. V. Riley, who in 1870 evolved the idea of using paris green to poison the potato beetle. This was a great success and rescued a valuable crop which was threatened with permanent destruction, as the insect spread over the country from its original home in the Southwest. Since that time the insect work has grown until about a thousand entomologists are employed, and a very extensive field of knowledge and technique has developed. Here it is especially evident that purely scientific investigation affords the foundation for practical results, and so the Bureau of Entomology includes men who devote their time to classifying and naming mosquitoes, flies, beetles, and other insects, to working out details of behavior and anatomy, and to studying breeding habits and immature stages, as well as those who are actively concerned in observing and checking serious infestations, wherever such may occur.

The native species are not alone concerned in the billion dollars worth of damage done annually by injurious insects. Some years ago the European gypsy moth and browntail moth became established and made a good start toward destroying the trees of New England; and their ravages have been held in check only through the constant efforts of experts and the expenditure of more than eight million dollars. At the present time the European corn borer, the larva of another moth, threatens to become a serious pest, a pest which would probably destroy the corn crop almost entirely if allowed to spread and multiply without attention from the economic entomologists. It was introduced into Massachusetts, probably in broom corn, in 1917 or a little earlier, and has spread as far as New Hampshire, Ontario, and Ohio. Poisoning is of little use, as the caterpillar lives inside the stalks of corn, dahlia, and other plants; so the methods of control, as deduced from scientific studies of the insect, consist chiefly of destroying the plant remains in which the species hibernates, and in preventing its further spread. If you are stopped on the road by an officer with a flashlight, it may not be a case of misplaced prohibition zeal; quite likely you will be asked if you have any vegetables or cut flowers in the car. The purpose of the inspection is to determine whether or not you are unwittingly a party to the spread of the corn borer into some hitherto uninfested locality.

Another foreign undesirable is the Japanese beetle, introduced into New Jersey about 1916. The adult insect feeds on leaves and fruit of many different kinds, doing great damage during June and July, while the grubs (young stages) live in the ground and eat grass roots, often injuring lawns. This species appears to find our climate suitable, for it multiplies enormously and is spreading gradually to neighboring localities. It is thus of great potential importance, and already an elaborate organization of government men is arrayed against it. Heavy poisoning gets some of the adults, and thorough cultivation of the soil may reduce the number of living grubs; but these measures are not very successful, since strong poison injures many plants and since a good deal of ground is not open to cultivation. In this case, again, stringent quarantine is being attempted, to prevent the spread of the pest, and great efforts are being made to discover or invent some improved means of control.

The reason why introduced species often become so important as destroyers of wealth is because they leave most of their natural enemies behind when they emigrate. Whether the animal is taken to a new region by human agency (on purpose or accidentally), or whether it migrates of its own accord, it may find the new region suitable in climate and other living conditions, while being devoid of such parasites, predators, and diseases as serve to keep its numbers down in its original area of distribution. Native species often multiply and become economically important because a suitable food plant is extensively cultivated by man, and thus life is made easy; but in such cases a more or less equivalent increase in natural enemies will usually occur. This is not so likely with introduced forms.

Realization of these ecological factors has led to a new method of control, now in active development. If the parasites have been left behind, why not search them out and bring them again into contact with their prey in the new region? It is at once evident that a good deal of strictly scientific zoological work is called for in carrying out this idea. The efforts of experts in classification, field work, breeding and rearing, embryology, and behavior, may be required in combination, and over long periods of time, before a foreign species can be discovered, introduced, reared in numbers, and established as an efficient enemy of the particular insect concerned. Several successful attempts of this sort have been made, such as the introduction of an Australian beetle into California to destroy a scale insect (an aborigine of Australia) which seemed sure, otherwise, to destroy the orange groves, and the use, in New England, of European foes of the gypsy moth. There are special laboratories of the Bureau of Entomology dedicated to this work, where experimentation is now being done on a great variety of foreign species, any one of which may prove to be of the greatest economic value, once the biological problems involved have been worked out.

It has been estimated that natural forces, including insect enemies, unfavorable seasons, and bacterial diseases, serve—on the average—to prevent about 90 per cent of the damage to agriculture that insects would do if unchecked. But it is the remaining 10 per cent that often makes the difference between the farmer's profit and loss. Thus, economic zoology is really concerned with providing the relatively small amount of artificial aid which is needed to supplement the normal amount of natural control. It is this situation that makes practically effective the measures of protection, which at first sight seem ridiculously inadequate in the face of the swarming billions of the enemy. It is indeed a narrow margin that exists between victory and defeat in this contest; and some of the experts most competent to judge believe that human welfare, even human existence, will, in the end, depend upon its outcome. Insects will very likely prove to be the most formidable contestant with man for the ultimate possession of the earth. As time goes on, all the resources of science will be needed in this strife, just as they are now employed in war between human enemies. Already aeroplanes are being used to spread poisonous dust over large areas of growing crops; poison gas is applied in confined spaces, such as warehouses, greenhouses, and trees covered with canvas; and the most recent advances in communication systems and automotive invention are employed. When crop production is increased by these means, human wealth is increased; and if the resultant fall in prices works hardship on the producers, we need not hold that thereby the work of economic zoology is rendered vain. Rather should economists be stimulated to find and cure the weaknesses of a social system that permits the increased production of wealth to work hardship on a large and absolutely essential group of the population.

We may now glance more briefly at some other branches of economic zoology, in which, likewise, the work is carried on in the belief that the conservation and improvement of animal life as a source of wealth is, in itself, justifiable, regardless of economic fallacies in the fields of distribution and finance. The ocean is a potentially inexhaustible source of wealth because of the vast numbers of food animals it can support with a minumum of human cultivation. Fishes, mollusks, crustaceans, and echinoderms swarm in it almost infinitely. Almost, but not quite. The Bureau of Fisheries owes its existence largely to the fact that oysters, lobsters, and certain valued food fishes have become greatly reduced in numbers (in accessible areas) through excessive catches, wasteful methods of handling, destruction at the breeding season, injury by the waste of large cities, and other preventable factors. It is clear that the problems indicated here involve scientific studies on habits, distribution, life histories, etc., to provide a reliable basis for practical laws and recommendations. Oceanic expeditions, requiring large expenditures of money and months or years of service by specialists, are carried out to determine the distribution and breeding places of various species; and government vessels patrol the coasts and survey nearby waters in minute detail. Other agencies

[309]

are similarly engaged in the study of animals which furnish milk, meat, eggs, silk, and honey. Current investigations of especial importance deal with the diseases which affect such economically valuable animals and with the heredi- tary factors which determine the qualities of various breeds, with the result that their productiveness is being contin- ually increased.

These examples, chosen from among many, will serve to illustrate how zoology is contributing to human welfare by increasing the real wealth that is available in the form of useful things. Many agricultural products are made abundant and cheap by improved methods of controlling destructive pests; and the value per head of domesticated animals is greatly increased by scientific breeding according to the principles of modern genetics. All such results depend ultimately upon purely zoological knowledge; and many of them have issued directly from specially directed research. Altogether, an enormous increase in wealth is represented; and if this increase is not felt at once by the ultimate consumer, it is the fault of our economic system, not of science.

IV

Given a satisfactory degree of health and wealth, the welfare of a human being—his chance for living the good life—still depends very largely upon the social relations in which he finds himself. And since man is primarily an animal species, his social relations are bound to be formed, in part, upon zoological principles. Perhaps the two most important relations are concerned with union for defense and with the family. These have a long zoological history; and we may consider them briefly as appropriate examples to illustrate our present point of view.

The social mode of life is, in many respects, of great advantage; chiefly, perhaps, because of the scope it affords for specialization. The group as a whole is largely responsible for defense against enemies, and so individuals—sometimes whole classes of individuals—are left free to devote themselves to getting food, building shelters, caring for the young, etc. This social division of labor tends to make the group efficient; and all the individuals benefit by the safe, stable, and complex mode of life which can often be attained in this way.

While the ants and the human species have been most successful in forming such a social organization, there are many other animals which have developed it in a lesser but still useful degree; and a study of the latter helps us in understanding and appreciating more complex social states as well as in inferring the probable course of their evolutionary history. Among the fishes, for example, group life takes the form of a loose aggregation or swarm, in which the chief advantage seems to be no more than the mere weight of numbers and the certainty that enemies cannot accomplish complete extermination. Its maintenance involves enormous reproductive activity, which exhausts most of the energy of the species; and there is little or no opportunity afforded for specialization among the members of the swarm. In the flocks or herds of various birds and mammals, certain individuals act as leaders, scouts, or lookouts, and so division of labor begins to appear. In certain insects, such as the ants, life follows complex and rigidly organized social patterns, the individual being almost completely subordinated in the interest of the group as a whole. Various castes may exist, devoted to certain duties and adapted to them not only in activity and skill but even in fixed, hereditary structure and instincts. Some species of ants have a military caste made up of members unable to do ordinary work; and some have a sexual or reproductive caste so specialized that its members cannot even feed themselves.

Human society is similar in many respects, but fundamentally different in psychology. In union there is strength, in animals or in human groups; and in both cases there is efficiency in specialization. But individual initiative and adaptability, as opposed to caste specialization, and almost the whole of that cumulative and changing social product called culture, can be found only where behavior is intelligent—that is, in human society. Instinctive behavior depends upon inherited patterns in the nervous system and, as practiced by the insects, can produce highly complex social relations which are eminently successful, if the individual is devoid of aspirations toward independence, and if environmental conditions remain substantially constant; while intelligent behavior depends largely upon learning and experience in the individual and, as practiced by man, can produce institutions equally complex but far more adaptable to varied conditions and, furthermore, distinguished by a historically developing culture of esthetic and technological character. Among the other vertebrates some traces of intelligence can, of course, be found, and these give zoological evidence of the way in which intelligent behavior evolved along its divergent and continuing way while instinctive behavior reached its early summit . . . and limit.

In both cases the advantages of social life have been fully demonstrated, and in the course of time, the social organisms have come to dominate the solitary types. Many of the latter, it is true, have escaped extinction on account of some special quality, such as extreme fecundity, wide latitude in food requirements, or exceptional adaptation to peculiar conditions; but the power of social solidarity has been a notable factor in natural selection, and goes far to explain the success of the ants and of man. The Darwinian principle, one of the great generalizations of biology, has been exemplified in the development of

[312]

human society, and there can be no doubt that it is still in force under civilization, where the weak fail in the face of unfavorable environments, no less surely than they do in the jungle, leaving to the strong both the earth and the accumulated treasures of human effort.

The family is clearly a zoological group in essence. It exists in more or less permanent form throughout much of the animal kingdom, even where no other form of social life is distinctly developed; and it depends fundamentally upon cytological and embryological processes which are essentially alike in all animal species above the Protozoa. Human welfare depends, to a formerly unsuspected degree, upon rational regulation of the family; and such regulation we now clearly see to be impossible on a superficial basis of social custom. The problems of birth control, of population increase, of delayed marriage, even of marital dissatisfaction and discord, in so far as they are soluble at all, can be capably attacked and understood only through the application of biological knowledge as well as sociological and humanitarian concepts.

But perhaps the most distinctively biological feature of the family is the inheritance of traits from one generation to another. Formerly, this passing on of characteristics from parent to offspring was a complete mystery, but its *modus operandi* is now clearly understood, as is explained in another chapter. We wish, in concluding this section, merely to point out how directly and poignantly the new science of genetics (based upon scientific research in zoology and botany) bears upon human welfare through the family.

The child defective in body or in mind, the "black sheep" that is supposed to appear in the best of families, the youth who goes to war and returns fit for nothing but a psychopathic hospital—all such, in a vast majority of cases, fail to cope with difficulties relatively harmless to

[313]

their fellows, or fail to succeed in favorable surroundings, for the simple reason that they began life under a fixed handicap. This handicap was not bestowed upon them because of exceptional ability and as a mark of honor (as in sports), but because their parents produced defective factors of inheritance which happened, by chance, to find expression in the unfortunates under consideration. This negation of human welfare is theoretically preventable, if persons who carry defective hereditary units relinquish the privilege of parenthood, and it can be minimized if the situation is fully recognized (as it is, for example, in certain schools and hospitals). To learn the facts and make sound scientific recommendations in this field is the task of eugenics. No one can say that this task is more than begun as yet; but already sterilization and segregation of defectives is a recognized social procedure within limited bounds, education is beginning to recognize inherent grades of ability, and students who get some knowledge of biology are aware of other desiderata in marriage than money and obvious personal qualities. Family counts, but in a new way.

V

We do not know that any animal other than man has an intellectual outlook or a personal philosophy of life, but we do know that an individual of the human species experiences his share of human welfare and contributes to that of others in a degree very largely determined by his mental attitude. What does zoology contribute to the formation of a good philosophy of life?

In the first place, zoological study, like that of any other science, fosters in the student a concern for facts and a tendency to regard facts objectively. This helps to counteract the common habit of cherishing ideas that are based upon imaginary wish-fulfillments rather than upon truth—

a habit which is an immemorial foe to human welfare. Some tinge of scientific detachment is surely essential to an adequate philosophy, if a reasonable contentment is to flow from it, for to view nature egocentrically is to court disappointment. Secondly, some definite knowledge of zoological and other scientific facts is a necessary part of the mental equipment of any one who cares to live comprehendingly in the modern world. It is apparently an unfortunate combination of fragmentary or vestigial knowledge of scientific fact—with a lingering and often unconscious yearning for the mystical certainties of a bygone day—that makes so pitifully inadequate the philosophy of certain moderns who are good writers and thus reach a wide audience. In his recent book, "The Conquest of Happiness," Bertrand Russell has devastating things to say about these vocal exemplars of maladjustment. A sound, though not necessarily extensive, experience of scientific work, such as is afforded by a good college course or two, will help to insulate the intelligent young against this form of injury to human welfare, and, at the same time, it will equip them to live sympathetically and safely—to feel at home—in a civilization which is distinctively the product of science.

There is another, and perhaps an equally important, way in which zoology contributes to the formation of an intelligent mental outlook. Evolutionary philosophy was developed from a condition of vague speculation through the efforts of many scientists, but it remains the especial glory of biology. Darwin's work made the evolution theory a living reality, and it inspired other zoologists and botanists to labor at the elucidation of multitudinous details, with the result that any one who is to gain a genuine knowledge of the principle must study the biological sciences. With this background it is possible to view, comprehendingly, modern philosophy, sociology,

[315]

religion, and other studies in which the idea of evolution is an essentially vitalizing element. In consequence of such a view, the mental horizon is tremendously enlarged, and an intellectual outlook is attained from which the shadows of superstition and irrational fear have departed. Nothing could contribute more directly to human welfare than a study which is capable of imparting this liberating philosophy.

And there still remain to be mentioned certain intellectual benefits afforded by the scientific study of animals which are of real value, though they may be thought lacking in high philosophic dignity. A first-hand acquaintance with earthworms, fruitflies, insects, and mice—gained in anatomical and genetic studies—removes forever the stupid attitudes of fear or disgust which ignorant people often take toward such creatures, and thus helps toward a rational appreciation of nature. More than this, when we have noted how a mere Protozoan, like Paramoecium, moves about in its miniature world in a way startlingly similar to our own movements of trial and error in exploring a dark room or in pursuing the scientific method; when we have observed identical modes of eye-defect inheritance in a fly and in ourselves; and when we have found the same bones, nerves, and muscles in an ape as exist in our own bodies, we come into permanent realization of our intimate relationship with the lower animals. General recognition of this relationship will lead naturally to public confidence in the practical value of animal experimentation and willing support of institutions where it is carried on, to the sure advancement of human welfare. Finally, it may be suggested that persons who have studied such natural processes as excretion and reproduction, in surveying the physiology of animals, are not likely to be unduly impressed with inane taboos handed on to them from an ignorant and squeamish past. This is a point of greater importance than it at first may seem, for the insistent inculcation of these taboos in early childhood is responsible not only for a foolishly artificial attitude in regard to matters of everyday physiology, but also for the building up in the growing individual of abnormal phobias and useless inhibitions that are often ruinous to welfare. Children who have parents with some biological training will be likely to escape this danger.

We hear, on all sides, that what the distraught inhabitants of our modern world most need is an adequate philosophy of life to rescue them from tumbling ignominiously into a vaguely defined abyss that seems to be yawning before them. Some saviors of society propose a retreat to fixed religious standards, humanistic or orthodox; others believe that a new economic scheme is what we need; still others call for fundamental changes in human nature, such as the assumption of disinterestedness. And there are the frank pessimists, already noticed, who should be for jumping in at once and thus putting an end to it. All such prophets seem unworthy of serious consideration to those whose philosophy of life is fortified by science. In this section we have seen how zoological knowledge, in particular, can afford some elements useful in the formation of an intellectual outlook adapted to the environment in which we perforce must live; and we who believe that it is possible for life to be reasonably happy and comfortable will not be surprised, when scientific knowledge in general becomes the basis for social ethics, if the consequent advancement of human welfare goes far beyond the present expectations of a reasonably guarded optimism.

Chapter XI

EFFORTS TO INCREASE FOOD RESOURCES

by Donald F. Jones

HISTORY has much to say about generals and battles. Its pages are filled with the deeds of emperors and kings, too seldom glorious. But the major factor in the growth of states and empires has been the origin and development of domesticated animals and cultivated plants. The United States has become a rich and powerful country, primarily because maize, the corn of the Indians, was so well adapted to the vast areas of tillable land that it laid the foundation of a prosperous agriculture. Canada developed an early maturing wheat, well fitted to northern soil and climate, and so opened a new domain for settlement. Ireland lost one and one-half million people by death and emigration when, for three successive years, the fungous disease Phytophthora ravaged the potato crop. After the discovery of a chemical spray that protected the plant from the fungus, the Irish prospered.

The plants and animals that nourish us, clothe us, and shelter us from the weather are so taken for granted that few stop to think how recently they became available. Only after long effort and many years of patient searching for useful plants and animals in every part of the world, only after their gradual improvement for man's best uses, did these products reach their present value in agriculture, commerce, and industry.

In "Ivanhoe" there is a banquet scene which gives a vivid picture of early days in twelfth century England.

The board groaned with pork and beef, with venison
pasties and roasted fowl, and ale and mead flowed freely;
but could we go back there, by some such miracle as Mark
Twain invoked in placing a Connecticut Yankee in King
Arthur's court, we should miss many things that are on
our everyday bill of fare. At that time purple-skinned
potatoes the size of marbles were being grown by the
Indians in North and South America, but no one in other
parts of the world had ever seen them. For thousands of
years Europeans ate turnips, carrots, and beets or did without vegetables of this type. Peas and cabbages there were,
but no knight in armor ever ate a tomato salad. Queen
Guinevere never tasted corn on the cob. Even to-day few
Englishmen know the delight of roasting ears. Can you
imagine a Thanksgiving dinner without turkey, without
cranberry sauce, and without pumpkin pie! No wonder
Thanksgiving is a more recent institution.

In those olden times there was plenty of smoke from the
oil-wick lamps and the open fireplaces but none of the
fragrant aroma from after-dinner coffee and cigars. And
we still call them "the good old days."

The ladies in our imaginary twelfth century excursion
party, after they had ceased to wonder at the windows
without glass, to say nothing about screens, and kitchens
with neither stoves nor plumbing, would soon miss many
of the beautiful flowers and shrubs that make the modern
home attractive. There would be no sweet peas on the
table or dahlias in the garden. Tulips and hyacinths and
many of the other spring flowering bulbs would be notice-
ably absent. Wild roses would be there in abundance but
no Pernet or Van Fleet creations. Lilacs, spiraeas and
deutzias, if grown at all, would not be represented by the
free-flowering and gloriously beautiful bushes of to-day.

Going into the orchard of our Saxon host we might
find some apples and pears, perhaps half the size of a

Delicious or a Bartlett, a few plums, cherries, and grapes; but the fruit lover would look in vain for peaches. Strawberries would be there, but only the little wild fruit that takes forever to pick. The grapes, too, though long cultivated in the Old World, would not be particularly fine in England. And even at their best the meaty, sweet European grapes do not have the flavor and juiciness of the native American grapes, so that many will agree with Longfellow in saying:

> Very good in its way is the Verzenay
> or the Sillery, soft and creamy,
> But Catawba wine has a taste more divine,
> more dulcet, delicious and dreamy.
> There grows no vine, by the haunted Rhine,
> by the Danube or Quadalquiver,
> Nor island or cape, that bears such a grape
> as grows by the beautiful River.

America contributed much to Europe; but there was an even exchange. In fact, one may trace the westward course of empire more clearly by the conquest of the soil and by the introduction of new plants and animals than by military triumphs. Following the paths of Cortes, Pizarro, Balboa, and others who killed and robbed the natives, came the Spanish missionaries to tell the Indians about the blessing of Christian civilization. They rode their burros throughout Mexico and what is now Florida, Texas, New Mexico, Arizona, and California. Here and there they built their monasteries and planted their grapevines, date palms, olive trees, orange groves, and alfalfa fields. These plant introductions did not thrive everywhere, but in many parts of the country they were the beginning of extensive and profitable industries.

While this development was taking place in the South and West, a similar pageant of colonization was taking place in the East and North. It began with a small sailing vessel that set out from Plymouth, England, stopping at Holland and landing on the Massachusetts shore in 1620. The passengers on the *Mayflower* are well known. What did they bring with them? Planning to make this new world their home, the Pilgrims did not come empty handed. Their ship was small, and no farm animals were brought on that first trip. But there were sacks of seed wheat and of barley; and packet upon packet of the seeds of other foodstuffs. In fact, practically the whole list of European garden vegetables came over with the early New England and Virginia settlers.

How would these crops grow in the new soil and unfamiliar climate? was the question the anxious settlers asked themselves. Would they ripen before the fall freezes? What new insects and blights awaited these plants upon which so much depended? Governor William Bradford tells us the story. The English grain, when planted that first anxious year, "came not to good, eather by ye badness of ye seed, or lateness of ye season, or both or some other defecte." Although the *Mayflower* passengers preferred the grains to which they were accustomed, they were forced by dire necessity to use an unfamiliar food—a strange-looking cereal that grew on a tall stalk and carried its seed on the side of the plant instead of at the top, as all grains were carried in Europe. Indian corn kept the members of the colony alive until they learned how to grow their Old World crops under the conditions of a new country.

Planted in the fall, English wheat and barley, adapted to a mild climate, could not withstand the cold of New England winters. When planted in the spring, it blasted and mildewed and would not ripen. Fortunately for the early settlers, grass grew almost as luxuriantly, if not so green, in the New World as in the Old; and the cattle and sheep brought over in the ships that soon followed the *Mayflower* thrived and multiplied. Their chickens and ducks found the meadows of Massachusetts as good a place to pick up a living as the fields of Devonshire. In the woods there was a larger fowl that the Indians prized. Carried back to Europe it was called the Cock of India at first, and later the turkey. So, to-day, North America's only contribution to the list of domesticated animals is named after a country in Asia Minor. But no more is the lordly turkey allowed to roam the fields and roost in tallest trees or windmills. He isn't even permitted to have his feet on the ground. Traveling about the country one sees these big birds in wire-bottomed pens. Bacteriologists have learned that the earthworms and other natural food of the turkey carry a deadly parasite that in the past has wiped out whole flocks. Such a simple procedure as keeping the birds off the ground has made turkey raising profitable in the part of the country where they once proudly strutted in the wild.

In the same country in which the Indians numbered less than ten million inhabitants, and then practiced a severe limitation of offspring, the white inhabitants now total a hundred and twenty million and are increasing at a rate that will double the present population in a half century. Where the Indians often went hungry, we now study books on how to reduce. Farm surpluses cause the government as much anxiety as the famines did in Egypt. In part this is due to better machinery for cultivation and transportation. Where the squaw with a clamshell hoe had to work hard to grow enough corn and beans to feed her own family, a man with a steel plow and cultivator can grow enough for two families. To-day, with power machinery, one farm family can produce enough to provide four or more families with an abundance of everything.

[322]

This development could never have taken place if the crops were not adapted to the varied soil and seasonal conditions in those regions where they are best grown. Some crops are better adapted to machine cultivation and others are preeminently fitted for hand cultivation. Although machinery has now been skillfully devised to plant, cultivate, and harvest the corn crop, all of these operations for thousands of years were done by hand. Wheat, on the other hand, has always necessitated machinery in some form. The seeds are too small to sow singly by hand and must be broadcast. This demands some form of raking the soil to cover the seeds. The plants need no cultivation or weeding, and when the crop is ready to gather, it has to be cut with some kind of knife, flailed with some kind of beater, and cleaned with some kind of blower. Through the ages man's inventive genius has been directed toward better means of performing these operations.

We think of the early explorers busily searching the earth's surface for gold and silver to make ornaments, and for copper and iron to make swords and armor. But one of the main uses for metal has always been to make food easier to produce. The hoe and sickle were first made of stone, then of bronze, and finally of iron and steel. With a little alteration in shape, the hoe became a spade and then a plow; the sickle became a scythe and then a cradle and finally a reaper. Where formerly it took one man a good part of the season to prepare the ground, plant the seed, and harvest and thresh an acre of wheat, the same thing can now be done in less than three hours of actual working time—plowing and planting in one operation, harvesting and threshing in another.

But what value would all this machinery have if the plants did not grow properly and ripen a useable crop? The New England settlers, as we have noted, could not use their English wheats, because they blasted and winter-killed; they were forced to use the hand-tilled grain of the Indians. Only when they brought wheat from the colder parts of Europe and Asia were they able to grow satisfactory crops. The apples and grapes from the Old World often failed in the New, and not until countless numbers of seedlings had been grown were fruits developed that could live through the coldest winters and bear edible fruit in the driest summers.

As the covered wagon went westward, many a fine piece of furniture and valued silverware was left behind, but not the seeds and plants that had been selected through so many trying years and upon which the settlers pinned their hopes in the wider fields and blacker soils of the West. Once again they were met with disappointment. The Bartlett pears and Baldwin apples could not survive the winters that were even colder than those in the East. Corn from east of the Alleghenies withered and died, scorched by the hot dry winds of the prairies. The wheat was devoured by grasshoppers, stung by the Hessian fly, sucked dry by plant lice, and rusted and blackened by fungous diseases. Even the flowers that grew with so little care along the seacoast and the inland valleys became discouraged and failed to bloom.

Again began the laborious process of searching for new crops that would stay green when the old ones withered. As more of the fields were plowed, the grasshoppers came in fewer numbers. And the newcomers gradually learned when to plant to avoid the flies, and how to produce varieties that would not rust or smut even in years when the old crops were badly damaged.

To take the place of the Baldwins and Greenings they found the Jonathan, Winesap, and Ben Davis apples. Red raspberries would not grow, so they found large-fruited plants of the wild black kind. Hardy plums were crossed with sweeter kinds, and a new fruit is now grown wherever plums are wanted. To take the place of the laurel and rhododendron about the house, coralberries and snowberries were taken from the pastures. Flowers were brought from the drier and warmer regions of the earth, and bloomed abundantly. Vegetable varieties were found that would do well on the various soils and in the dry air.

The early settlers first used pigs and poultry, but the larger farm animals were brought over as soon as their means permitted. The ocean passage required about three months, and this was very trying for animals in confinement, half of a shipment often dying en route. Food and water for the voyage were no small items. As the farmers became more prosperous, they brought from England the finest Shorthorn and Hereford cattle that money could buy, the best Shropshire and Southdown sheep, and Yorkshire and Berkshire pigs. From France came Percheron horses, and from Germany came Holstein cows. All of these breeds had been centuries in the making, but many of them have since been brought to their finest development in the valleys of the Ohio and the Missouri.

The possibilities from new animal importations are by no means exhausted. Reindeer from Northern Europe have recently been established in Alaska and are valuable producers of meat. A pigmy hippopotamus from Liberia has been suggested as a meat animal for southern swamps. Ostriches are grown in the Southwest. Zebus from India, crossed with native cattle in the South, thrive where the older breeds are subject to disease. In return for all these valuable animals from Europe, America has sent back little in kind. The native bison has possibilities as a meat producer, but up to the present time has never been used profitably in any part of the world. The mountain sheep and prairie hen were large enough to be used, but the animals of this type already tamed were so well developed

[325]

and so tractable that no one had the patience to spend time and effort in domesticating the former. In South America the llama and alpaca are used locally but do not seem to have much value for other parts of the world. The only animals to be taken back to Europe and used were the guinea pig from the southern continent and the turkey from the north—and neither of these is important. Our plants have been more useful. Maize, originating in Central America, is now grown in Southern Europe and in parts of Asia, Africa, and Australia.

The original home of maize was wrongly identified because the names under which it was first grown in Europe indicated an eastern origin. The word maize (mays) is American, as it comes from the Mayan tribe in Central America, but this name was not used at first. In English, the term Indian corn suggests India as its home. In French, *blé de Turquie*, or Turkish wheat, indicates Asia Minor as its birthplace. But since maize is not wheat, one may also suppose it is not Turkish. Many plants are thus wrongly named. English walnuts did not originate in England, nor are they extensively grown there. Oriental persimmons are called date plums, although they are not related to dates, nor are they plums. The Jerusalem artichoke shows how misleading a name can be. The edible part of this plant is the root and not the blossom bud of the true artichoke. Nor did it come from Jerusalem; it is, in fact, the only cultivated plant native to the temperate region of North America. Its name seems to be merely a corruption of the Italian *girasole* meaning sunflower. As one botanist puts it, "we thus see how much worse a double name is, since if it be single only one mistake is possible."

The names which have been given to corn in Europe simply indicate a foreign origin. Maize was called Roman corn in Central Europe, Sicilian corn in Italy, and Turkish corn in Sicily. The Turks called it Egyptian corn, while in Egypt it passed as Syrian dourra. No word for this grain has been found in the ancient European and Asiatic languages. An ear of corn was found entombed with an Egyptian mummy, but it is certain that this was the work of a modern impostor, since the seeds grew when planted. If maize had been grown in Egypt by the Egyptians, it surely would have been pictured in their paintings and sculptures, along with their other valuable plants.

On the other hand, there is every evidence to show that corn was the principal food of the natives of America and had been since remote times. It was named in every tribal language and had a prominent place in their religious ceremonies. Seeds have been found in the remains of mound builders and of cliff dwellers. The tombs of the Incas and the temples of the Aztecs held the sacred grain, just as in Egypt the sepulchers and sanctuaries were made the depositories of wheat and barley.

Other important plants to go from America are the potato, sweet potato, tomato, and tobacco, nearly all members of one plant family. Other vegetables of less importance are grown, but none to the same extent. The cinchona tree, from which quinine is made, comes from South America; the eucalyptus tree is an important introduction from Australia; and the rubber tree from Brazil has had a profound influence on the development of the Malay archipelago and some of the East India islands. It is now being tried extensively in Liberia and the Philippines, and promises to become one of the most important trees grown. Plants in a long list are used for drugs, flavors, perfumes, oils, fibers, rubbers, resins, gums, and waxes in all parts of the world. Many of these are grown in regions far removed from their place of origin. Systematic botanists, exploring the earth on collecting trips, studying the plants in herbaria, and growing them in botanic gardens, have pointed out possibilities of usefulness in many of

[327]

these species. Plant physiologists with the newer knowledge of biochemistry have many things still in store for us that will add to the value and beauty of plant products.

Much of the best effort of science is directed toward the reduction of the loss from insect pests and fungous diseases. Along with the immigrants from across the ocean have come many stowaways that we should have preferred to have had stay at home. When the Hessian troops were hired by King George to fight the colonists, they brought with them straw from Germany to feed their horses. In this straw, it is now supposed with good reason, were some oval-shaped bodies about the size of a flaxseed. When spring came, these inert cocoons took on life and hatched into clear-winged flies whose progeny have done more damage than all the armies that have ever marched on American soil. The Hessian fly is one of the serious pests of wheat that has been held somewhat in check because entomologists, studying its life history, learned that the flies cease to lay eggs after the first frost in the fall. The planting of winter wheat is now timed, as nearly as is possible, so that the plants will not be up until after frost.

One can not ride far these days without being stopped by inspectors looking for plants that may be carrying an unsuspected foe. The European corn borer, the Japanese beetle, and the Mexican bean beetle are now spreading across the country with a persistence that seems to be impossible to check, just as in the past the Colorado bettle, whose food was formerly the wild nettle, took to the potato and spread wherever potatoes were grown in this country. The cotton weevil, the pink bollworm, the codling moth, and the chestnut-blight disease have all entered the country without our consent. Most of our serious plant and animal diseases have come with their hosts when the latter were introduced, but some have found other plants

[328]

better to their liking, and some have stayed at home for a while and immigrated later. The barberry can not be planted where wheat is grown because it is a propagator of a rust disease that places an annual tax of an enormous sum. Currant and gooseberry bushes have the same relation to a disease of pine trees and are ruthlessly dug out where the pine trees are first in importance.

Although loss can be prevented by proper cultural practices and timely applications of chemicals, how much less expensive and how much more satisfactory it is to have plants and animals that are immune to disease and insect injury. The most notable cases of this kind are the rust-resistant wheats and the mosaic-free sugar cane. Certain kinds of tobacco will grow vigorously in soils where other varieties are stunted by root-rot. Cabbages have been developed that will produce an abundant crop alongside fields carrying varieties that are a complete failure. Cowpeas that harbor nematodes in their roots can make no satisfactory growth, but some plants are so constituted that these parasitic worms can not enter them. American grapes, through many centuries of association with root lice, have become immune to injury from their action. Not so the European grapes, that had never known this pest until the early explorers brought back vines from America. Little did they realize what a frightful menace they were bringing to the Old World vineyards. Only by bringing over American vines to use as root stocks and by developing methods of spraying, have Europeans been able to continue grape culture. On their new roots the vines yield more than ever before.

Few people realize how fast the vegetables, fruits, and flowers are changing. The wheat that goes to the modern steel mill in railroad car, canal barge, and ocean ship is not from the same varieties that the ox cart and saddle pack carried to the stone gristmill a few generations back. The

[329]

housewife driving to Faneuil Hall market in 1850 with her horse and buggy would find no Green Mountain or Irish Cobbler potatoes, no McIntosh or Delicious apples, no Elberta or Carman peaches, no Concord or Niagara grapes, no Howard or Chesapeake strawberries. Golden Bantam sweet corn and Iceberg lettuce did not exist, and grapefruit was never seen.

What will the great-granddaughter of this same woman have to serve in 1950? It is certain that twenty years from now will see some notable changes in fruits and vegetables. Already lettuce, fresh peas, beans, and tomatoes are available nearly every month in the year. Instead of being grown on nearby farms, they come from Florida, California, Mexico, and Africa. New varieties are being developed that will grow better and yield more in the new localities, that will stand long shipment and be better colored and more attractive when put on display. Already early and productive stringless beans are available, so that there is no excuse for a stringy bean on any market. The Cortland apple hangs on to the tree, is well colored and productive, and has a better flavor than its McIntosh seed parent in spite of the fact that its pollen parent was the Ben Davis. The Mikado and Golden Jubilee peaches are new names with which the housewife will soon become familiar. Early Sunshine and Whipple sweet corn are replacing Golden Bantam because they are earlier and handsomer. Sheridan grapes have a richer flavor than the old Concord, and the Portland ripens days ahead of any other grape. Blueberries as big as cranberries can now be grown in any garden where the soil and climate are suitable. Not a single variety of major importance eighty years ago is a leading variety at the present time. Many of the best fruits, flowers, and vegetables of the present day were then unknown. How were these new plants produced? In some cases they were merely chance discoveries.

[330]

A schoolmaster by the name of Stair went strolling around the countryside in England about a century ago and came upon a pear tree that attracted his attention on account of its handsome crop of smooth, bright yellow fruit. How this tree came to be where it was, nobody knew. Perhaps a squirrel on his way to his cache had dropped a seed, or a boy may have thrown a core over the fence. Anyway, there it was, uncultivated and unpruned, its branches lopped by grazing livestock; yet it bore fruit so sweet and tender that Stair decided it was worth cultivating. At the proper time scions were cut and grafted into his orchard. In a few years these bore fruit. The attention of a neighboring nurseryman was drawn to this new pear. His name was William. Soon the new variety, known as the Williams pear, was distributed over the British Islands wherever pears were grown.

French and Belgian nurserymen were keenly interested in pears, but thought that another name would be more suitable; so when the Williams pear crossed the channel, it was renamed Le Bon Chrétien, and by this name it is now well known in Europe and Australia. Sooner or later such a good fruit was sure to come to America but, by the compensation of fate, its French name was lost on the way. It did so well in a garden at Roxbury, Massachusetts, that a man by the name of Bartlett, who had bought the place, thought that it should be made generally available; and, as he was not averse to having his own name attached to such a good fruit, this pear has since been known everywhere in the United States as the Bartlett. The trainloads of canned pears from California are mostly Bartletts, and four out of every five boxes sold on the fruit stand are from trees that are lineal descendants by vegetative propagation from the original tree that schoolmaster Stair found so long ago in England.

Near Newport, Rhode Island, Captain Green kept an inn that was famous in Colonial times for its apple pies. One of the important factors in the excellence of these culinary delights was a tree that grew in the seedling orchard nearby. As an especial favor Captain Green gave his guests scions from this tree. Spread around New England, Green's Inn apple became famous, and to-day the Rhode Island Greening is one of the "New England Seven" varieties of apples recommended for commercial planting.

In this way have come many varieties of fruit and quite a long list of vegetables and flowers. Conscious selection has played an important part, for man, since earliest times, has recognized the importance of heredity; and, while often disappointed, he has learned that, in a general way, "like produces like" and "to produce the best, it is necessary to breed from the best."

Seventy years ago Balzac wrote that "heredity is a sort of maze in which science loses itself," and he was right. Since that time biologists have learned much about inheritance, and mostly from one man, an Austrian monk. Gregor Mendel, abbot of Brünn, showed how it was possible to take a dwarf green-seeded pea and a tall yellow-seeded pea and by cross-fertilization to produce a dwarf yellow and a tall green pea. Starting with two old varieties, he now had two new varieties. Some of these new plants bred true, and some did not. Mendel showed what was to be expected from a cross of this kind in the following generations, what was the best procedure to bring together— in the same individual—characters that had existed separately before, and what to do to make these varieties reproduce true to type. From this modest start, workers in scientific laboratories have extended knowledge of heredity to many animals and plants. They can tell why it is that certain characters disappear in one generation and reappear in the next, why it is that the offspring have qualities that the parents did not exhibit, why some individuals are weak and sterile and easy victims of disease, and why others are strong and virile.

Even before Mendel's time plant and animal breeders had learned that crossing different varieties or different breeds made new and more desirable combinations of characters, and that in some cases entirely new traits were obtained. But Mendel's rules of heredity made this enterprise more purposeful and more effective, eliminated much wasted effort, and helped to avoid the discouragement that comes from trying to do impossible things. Already new wheats are spreading over the plains from Texas to Saskatchewan. Trainloads of crisp firm-headed lettuce are coming from California that grew in fields where disease had formerly caused sad losses. Go to any eastern market in the shipping season and you will find the stores piled high with crates of Marglobe tomatoes. New grapes, that have come from crossing the older varieties and selecting plants that ripen earlier, bear bigger bunches, and have a flavor that the Concords and Niagaras never have, are listed by nearly every nurseryman.

New York State is well adapted for growing grapes. The experiment station at Geneva—under the leadership of Beach, then Hedrick, and later Wellington—has long been interested in the production of better varieties. Some of the most successful developments are the Portland, Ontario, Brocton, and Sheridan grapes. All of these varieties are earlier, sweeter, or better flavored than the older varieties of the same season. As grown in Connecticut, Portland is the first grape to ripen, producing handsome bunches of well-flavored fruit. It combines many of the good qualities of the black Champion and the red Lutie, its parents. Unlike either, Portland is a green grape. On account of its early maturity, Portland may be grown where heretofore no variety ripened satisfactorily.

[333]

Ontario is another early green grape, not so attractive in appearance, but sweeter than Portland—better even than Winchell or Diamond, its two quality parents, and earlier than either. It is a welcome addition to the list of grapes for the home garden and is being widely planted.

The first grape seriously to rival Concord was produced at the Geneva station. This is the handsome Sheridan that gets its sturdy vine and rich luscious flavor from the old Herbert and its ability to yield abundantly from the well-known Worden. The large compact bunches of dark blue fruit, hanging in immense numbers, are a sight worth seeing. Sheridan was first named in 1920 and was distributed the following year. It has been tested in many places and, while it has some faults, these can be forgiven since there are so many good qualities to commend it.

No mention of the Geneva grapes is complete without naming Brocton. If one can imagine the juicy tenderness of Brighton, the sweetness of Diamond, and the flavor of Winchell, all combined in one fruit, he can form some idea of this quality grape. It is a tender fruit and needs extra care for its growth and can not be shipped far to market, but it is worth all the attention that it needs.

These are not all of the new grapes that have come from Geneva, by any means, and more are promised. Nor have the other fruits been neglected. One of the earliest productions to win the attention of fruit growers was the Cortland apple. Studying the list of varieties commonly grown in New York State, Beach saw in the McIntosh an apple of much promise but one with several serious deficiencies. When rightly grown the McIntosh is a beautiful fruit, juicy and well flavored. But it ripens too early for winter markets, and the first strong wind puts much of the best fruit on the ground. Often the trees fail to bear as full a crop as many other apples do.

Out in the West there is an apple that seldom fails to bear a heavy crop. Heat and drought leave it unharmed. The fruit hangs on in spite of prairie winds. In many places this Ben Davis is the only apple that can be successfully grown. To Beach, the Ben Davis seemed to have all of the qualities that the McIntosh lacked. In 1898 Ben Davis buds were emasculated and covered with bags to keep out the bees. A few days later, tiny pollen grains carefully gathered from McIntosh trees were placed on the Ben Davis stigmas. From the resulting fruit, the black seeds were taken and buried in moist sand over winter. Planted in the spring the young seedlings were soon ready to set in the trial rows. The Cortland tree at one time was just one among hundreds of similar trees, without a name or anything else to single it out from the many other off-spring of the same parents or from those of other parents. No plant wizard saw the Cortland in its awkward stage and said, "Here is a tree that will be grown by the millions." All it had was a metal tag to show that the seed came from a Ben Davis tree and had been fertilized with pollen from a McIntosh.

There are some varieties of apples that bear a basket of fruit when three or four years old; but most varieties take from five to eight years before producing any fruit at all, and some of the finest varieties must have ten years or more. This Cortland tree bore its first fruit in 1906, nine years after the first pollination was made. At first it attracted little attention, but by 1915 it had produced regular crops of such excellent fruit that it was considered to be the finest of the whole collection. After nine years of careful observation, the Cortland was considered worthy of testing by fruit growers.

On the desk before me is a Cortland apple grown in Litchfield County, Connecticut. Its rich red color tempts me. Its fragrant aroma is the inspiration for these remarks;

[335]

and if they seem to the reader to be over enthusiastic, I invite him to share with me the pleasure of sampling some of its fruit. When well grown it is beautifully colored, making it one of our most attractive apples. But fruit growers like it most because it hangs on to the trees until the fruit can be picked. When the McIntosh is beginning to shrivel and to lose its piquant flavor, the Cortland is just coming into its prime. It is an early winter apple, while the McIntosh is at its best only in the fall. The flesh is as white as the Snow apple—the grandfather, probably, of all the McIntosh family. Though it has a flavor much like the McIntosh, it is richer and sweeter and with less aroma, but more firm and just as juicy.

In the same way, hardy pears have been produced at the Iowa station for the upper Mississippi Valley where, up to the present time, the trees have badly winterkilled. Plums and grapes are also grown in Dakota and Minnesota where formerly little good fruit of this type was available.

Vegetable crops have received as much attention as fruits. Although grown for thousands of years in South America, where it was first planted by man, and in North America before the coming of the European settlers, the potato did not begin to assume its present form until the middle of the past century. Early Rose was the first variety of modern type to be generally cultivated. The history of Early Rose begins with the Reverend Chauncey Goodrich of Utica, New York. Goodrich had the idea that the potato, as a result of long continued vegetative propagation, had become weakened and was no longer able to withstand the disease that was then causing so much damage. Through the American Consul at Panama he arranged to have a number of different kinds of South American potatoes sent to him from which he planned to grow seedlings. One of the importations he named Rough Purple Chili, from its appearance and the place where it had been grown. Flowers formed; and from naturally pollinated seed balls he grew several thousand seedlings and in 1853 selected one for propagation that he called Garnet Chili. This has been grown up to recent times.

Garnet Chili was grown by Albert Bresee over in Hubbardton, Vermont, and from the plants he raised came the seedling that developed into the Early Rose. First grown in 1861, it gathered local esteem until in 1868 it brought a record price of one thousand dollars—its originator being awarded a medal by the Massachusetts Horticultural Society. This early maturing, pinkish potato was one of the first really important commercial varieties originated in the United States and has been grown at one time or another in every potato section of the country.

Early Rose has been remarkable, not only for its own good qualities but also for the many valuable varieties originated from it. In 1871 Luther Burbank gathered a single seed ball from an Early Rose plant growing on his farm at Lunenburg, Massachusetts, and planted the seeds the following spring. From the twenty-three seedlings thus raised came the Burbank potato that, at one time, was an important variety in this country and is still grown in a somewhat different form in the West. Many other seedlings were grown by Bresee, and some achieved great popularity. Some were introduced, before their originator intended them to be and to no profit to him, from tubers dug out and stolen from his trial plots. But not until ten years after the advent of the Early Rose did another variety appear that offered it serious competition. This was the Early Ohio, first generation offspring of Rose with an unknown pollen parent, the seed balls being fertilized naturally. This fine potato originated with Alfred Reese and has had first importance in the Red River Valley of Minnesota and North Dakota and the Kaw Valley of Kansas. Another and equally successful child of Early

[337]

Rose is the Irish Cobbler. Little is known about its origin, but this variety, with Green Mountain and Rural New-Yorker, make up the bulk of the commercial crop in the East. Green Mountain, along with numerous other varieties that are not now grown, resulted from the attention of O. H. Alexander of Charlotte, Vermont. The parentage is unknown.

The potato-breeding efforts of E. S. Carman deserve particular mention. This former editor of the Rural New-Yorker advertized in his paper for potato seed and, in response to his appeal, received quantities from many parts of this country and from Europe. From this collection he grew many seedlings which he attempted to interpollinate by hand. He had little success, but fortunately on some plants seed balls set naturally, and from these he grew the plants from among which were selected Rural New-Yorker, Carman, and Sir Walter Raleigh. All of these are important varieties. The first is grown by the millions of bushels in New York, Michigan, Wisconsin, Iowa, and Minnesota.

Within recent years little interest has been taken in the production of new varieties of potatoes, owing largely to virus disease. This insidious infection is widespread but little understood. It is less prevalent in northern regions having a short, cool growing season. For this reason nearly all seed tubers are grown in the north and shipped each year to the main potato growing districts all the way down to Florida and Texas. This is a tremendously wasteful procedure. It would be better in many ways to have the seed tubers grown locally. This would not only save the expense of hauling but would also make possible the production of new varieties better adapted to each locality.

Recently the United States Department of Agriculture has announced a new virus-resistant potato from the breeding grounds at Presque Isle, Maine. This potato has desira-

[338]

ble qualities of shape and yield, and, if it proves to be free from disease, it should be a most valuable production not only for its own culinary qualities but also for the production of other varieties better suited to diverse soils.

We have read much in newspapers, magazines, and books about Luther Burbank's creations in flowers and other plants, but how many of his productions can be found listed to-day in the seed, bulb, and plant catalogs, those final arbiters of lasting merit? The many beautiful flowers that are pictured in profusion and grown all over the country have been the result of painstaking and diligent effort by those who are for the most part unknown, men working in a quiet way in many parts of the world, in places that the newspaper reporters have never heard about, men whose fingers are more used to tweezers and camel's hair brushes than to the typewriter or fountain pen.

How many know where the Crimson Rambler rose started its rovings? Is the Silver Moon a chance-sown seedling? If not, who is responsible for its development? Although many of our best apples and peaches have come from seed dropped along the highways and byways, this is not true of many of our prettiest flowers and is particularly untrue of the rose. "Rose" is the only word that can be recognized in nearly all languages of the world. The plant exists in many species and grows wild everywhere.

In 1927 the American Rose Society asked the expert rose growers, those who had had long experience with many varieties, to name the best climbing roses. The seven chosen are necessarily of wide adaptability and, for that reason, may be excelled by others in particular places, but they are undoubtedly the ones that have given the most satisfaction and delight to the greatest numbers.

First on the list is the Walter Van Fleet rose, originated in 1910 by a worker in the United States Department of

[339]

Agriculture. Three other varieties of his origination are on this same list: American Pillar, Silver Moon, and Mary Wallace. Paul's Scarlet Climber is second on the list and is the only variety from outside the United States. The once popular Crimson Rambler—an introduction from Japan, whose early history is unknown—is giving way to flowers more nearly like the bush roses.

The man who has had probably the greatest influence on the development of the rose in recent years has been Jules Pernet-Ducher of Lyons, France. For forty years this skilled hybridist has been producing new and better roses, and he is the founder of a new class called the Pernetiana group. By crossing the hybrid perpetual, Antoine Ducher, with the Persian Yellow, he obtained Soleil d'Or and Austrian Copper. From these have come some of the loveliest of roses: Souvenir de Claudius Pernet, Souvenir de Georges Pernet, Madame Edouard Herriot, Mrs. Beckwith, Angèle Pernet, Lyon, Ville de Paris, Président Cherioux, Willomere, and Severine.

Of the fourteen Gold Medals awarded in the Bagatelle Rose Trials for roses originating in France, twelve have been won by Pernet-Ducher. For varieties of foreign origin, five have gone to Alex Dickson of Ireland, three to Leenders and Company of Holland, two to Howard and Smith, and one to E. G. Hill in the United States.

The way in which new roses are developed is illustrated by the two following incidents. Van Fleet first crossed an old tea rose known as Devoniensis with the Memorial rose. One of the crossbred seedlings seemed to have some merit, but he was not satisfied with it. Emasculated flowers were then pollinated by the Cherokee rose, a naturalized species from China that runs wild in the South. A seedling from this parentage was unexpectedly more hardy than any of its ancestors and, on account of its porcelain-white, saucer-shaped flowers, was appropriately called Silver Moon. This lovely rose is brightening homes and gracing gardens all over the United States.

One of the first large-flowered, hardy climbers that popularized this class of roses was Christine Wright. This variety came from one of four fertile seeds grown on a climbing American Beauty flower crossed with Caroline Testout by James A. Farrell at West Chester, Pennsylvania. One of the other four seeds produced Purity. To get such good roses from so few seeds is an event that is not likely to happen often. Other rose breeders have had to work harder for their successes. At Richmond, Indiana, E. G. Hill grew five thousand seedlings at one time. All but sixty were discarded as soon as they flowered; of these, fifteen were selected for further trial and only two were named and distributed. But these two were Premier and Columbia, roses that stand among the first in the greenhouse class and are widely grown outdoors.

Plants, being fixed in one location, must be adjusted to the physical and chemical nature of the soil and so regulated that they will ripen properly in the available growing season, even in the most unfavorable years. For these reasons the genetic constitution of plants will always demand a great deal of attention. Animals can be moved anywhere. If the weather is too cold, they can be housed. If their feed is not suitable, it can be changed. Livestock that does well in one part of the world often does equally well in another. Therefore farm animals do not need the attention of the breeders so much as plants. On the other hand, the individual animal is worth much more than the individual plant. The good and bad qualities of a domestic animal are more apparent, and the reproduction of animals is more easily controlled. The breeding of animals has always engaged the attention of farmers from the very first.

Early Egyptian carvings show that the cattle of the Pharaohs had many of the desirable qualities of the beef breeds of to-day. From similar evidence, we know Assyrian horses had the size and conformation of many of our best breeds. To a certain extent these idealized creatures existed only in the imagination of the artist, but they must have been suggested by the livestock in the fields and, judging from these art relics, the ideals of ancient herdsmen were much the same as those held to-day.

History does not go back far enough to tell about the beginnings of animal husbandry. The great differences between domesticated animals and their nearest wild relatives indicate the long time that they have been cared for in confinement. Agriculture was the principal form of livelihood before machines were invented; and, while there has been much improvement in mechanical invention, animal breeding proceeds in much the same way it did four thousand years ago. Belief in maternal impressions, mentioned in Genesis, is still often held, along with many other opinions that have no foundation in fact. The changes that have been made in animals, adapting them to man's better uses, are the result of the application of the simple formula: "select the best for breeding stock."

The first notable advance in animal breeding methods occurred about the middle of the eighteenth century in England. The red, black, and pied, short-horned cattle of northeastern England, interbred with cattle from Holland, became famous all over the British Islands. Certain individuals were exhibited in all of the important cattle raising districts. They came to be known by name. Farmers eagerly bought calves from these celebrated cows and proudly traced the pedigrees of their herd sires back to this foundation stock. From this source developed the system of recording pedigrees, and in 1822 the first English herd book was published.

Then followed the realization that the value of an animal for breeding purposes consisted in its ability to transmit good qualities rather than in its own appearance. Some of the finest individuals were failures as breeders of desirable offspring, whereas other individuals of poor appearance were the progenitors of splendid specimens of the animal breeder's art. Back of the poor breeders was usually noted some inferior stock, while the good transmitters usually came from a long line of good ancestry.

More recently it has been realized that good ancestry alone is not enough. In spite of illustrious forebears there are too many "black sheep" in animal families. One way to avoid undesirable offspring, science has learned, is to use tested sires. If high egg-laying hens are wanted, use cocks that are known to beget hens that lay many eggs. If the milk yield is to be increased, it is necessary to use a bull that has the ability to augment the milk yield of his daughters over the yield of their dams. Basing selection on progeny performance rather than upon the appearance of ancestors was the next forward step in animal breeding.

Although the best of the purebreds have desired qualities, they are not so generally used in everyday farm production as might reasonably be supposed. In 1920 the census reported less than 1 per cent of the horses to be purebred, 3 per cent of the cattle, between 1 and 2 per cent of the sheep, and $3\frac{1}{2}$ per cent of the swine. Less than 4 per cent of all livestock in the country are purebred and registered. According to Wentworth, little more than half of all animals have sufficient characteristics of the recognized breeds to be classed as grades from registered stock. Why is there this apparent lack of appreciation of the value of selected and registered livestock? Wentworth says:

"Commercially, we . . . seem to have reached certain limits beyond which the methods of pure breeding are not practicable. There is no question as to the improvement that a purebred beef animal . . . shows over a grade

[343]

animal . . . The characteristic differences are definite and well understood. Nevertheless, production costs have increased to such a point that the refinements of the pure-bred in beef conformation and finish subject themselves unfavorably to the law of diminishing returns, since consumers do not pay the price for fancy meats that ordinarily warrants their production. According to present methods of breeding, stimulation of demand for meats and further improvement of market stock require additional use of expensively produced purebreds. Hence we have arrived at the apparent *impasse* of requiring better breeding at lower price levels, whereas under practical conditions we can secure better breeding only through higher costs."[1]

It is becoming clearer and clearer that one of the best ways of using the inherent value of purebred livestock lies in a more extensive use of crossbreeding. The mule is the oldest example of a crossbred animal. This hybrid combination of the horse and the ass has long been appreciated, but, since it is physiologically sterile, it can never be used for breeding purposes. The blue-gray cattle of Scotland are famous. Their distinctive color comes from the genetic composition of black pelt from the Aberdeen-Angus, or Galloway, combined with the white of the Shorthorn. The quick growth and uniformly good meat quality of these mixed breeds are due to the phenomenon of hybrid vigor, commonly associated with the crossing of somewhat unrelated forms. Western sheep in the United States are almost exclusively crossbreds. Swine from Poland-China and Chester White combinations or Duroc-Jersey and York-shire matings are commonly seen in the Corn Belt. Since they can not be used for breeding purposes, there must be good reason for their production.

[1] *Record of Proceedings, American Society of Animal Production, Annual Meetings,* 1925–1926, pp. 42–43.

The most extensive use of crossbreeding is made in Denmark. Danish bacon has an enviable reputation. It is sold in England in competition with the best from British farms. About one-third of this bacon from Denmark is produced by mating the English Yorkshire with the Danish Landrace. Purebreds only are used. Each farmer buys his breeding stock from breeders who are approved by the government. While English and American breeders have spent their principal efforts in producing animals for the show ring, Danish farmers have developed a system whereby a pound of pork is produced from a little more than three pounds of feed. Careful tests of this have been made at the leading swine stations. At Elsesminde the Landrace has averaged 3.79 pounds of feed for 1 pound of gain and the Yorkshire 3.69 pounds. The crossbreds have averaged 3.23 pounds or 13.6 per cent less.

In the *Breeder's Gazette* there is a report by Casement giving the results obtained on his farm where Duroc-Jersey and Poland-China sows were both mated with Yorkshire boars. Many such cross matings were made and hundreds of pigs raised. The crossbred pigs are compared with purebred litters from other sows of the same breeds as the dams. There were, in round numbers, 10 per cent fewer pigs dead at birth, 30 per cent more alive at time of weaning, and 5 per cent more raised to market, of the crossbreds than of the purebreds. No figures are given for Yorkshire purebreds, so that the figures are not directly comparable to the Danish results; but they indicate clearly that there is value from crossbreeding in viability and longevity as well as in economy of gain.

The most critical evidence for hybrid vigor in domestic animals is furnished by Roberts and Laible at the Illinois Experiment Station. A purebred Duroc-Jersey sow was mated in the same period of heat to a purebred boar of the same breed and also to a purebred Poland-China. Six of the pigs in the litter from this double mating were pure-bred and four were crossed, as could be easily told by their color. All were born on the same day, nurtured by the same mother, and fed the entire time on exactly the same rations. The crossbred pigs weighed 3.75 pounds each at birth, the purebreds 3.23. After six months, the only two females of the six purebreds that were alive weighed together 371 pounds, while the two crossbred females weighed 456 pounds, making a total difference for the two of 85 pounds. The two crossbred males weighed 485 pounds.

Because crossbreds do so well, there is always the danger that farmers will use these fine-appearing animals for breeding. If this were generally practiced, the pure breeds would soon disappear as such. The reasons why crossbred animals are not desirable for continued reproduction should be thoroughly understood. The dominance of favorable growth factors is in the first generation, where the parental races are distinct but uniform in their genetic constitution. In later generations Mendelian recombination permits the reappearance of weaknesses and undesirable qualities that are overcome by crossing. Many of the progeny of a crossbred wool and mutton sheep will have the inferior fibre and light clip of one parental race together with the lean body and bony carcass of the other. The breeding together of Berkshire and Tamworth or Poland-China and Chester White crossbreds brings forth four-legged creatures that are inferior in size and vigor to the poorest specimens in either purebred.

"Registered livestock has been rigidly selected towards definite standards. Each breed has certain qualities which make it valued above all others. These qualities are built up to a higher degree than those of any large number of animals resulting from any other system of breeding. The pedigree record insures that they will come fairly true to that type. But purebreds, when compared with the better classes of non-pedigreed livestock, are often somewhat lacking in vigor, rapidity of growth, and fecundity, three very important considerations from the standpoint of the man on the farm who is making his living by producing meat and milk, wool and eggs. The very qualities which close mating unavoidably brings in are counteracted by crossing. Prolificacy, hardiness, and rapid growth are the outstanding features of crossbreds. In hybrid vigor there is a very real help towards increased production. Why not use it?

"There are several reasons why crossing is not more generally practiced than it is. In the first place, the breeders and sellers of purebred livestock discourage crossing. Nearly all writers on animal breeding see so clearly the absolute necessity of maintaining the invaluable pure breeds of livestock which we now have, and of making them still better, that they minimize the real benefit to be derived from crossing the breeds for production purposes. They hesitate to say anything which to them looks like breaking down that system which has had everything to do with building up the livestock industry to the high plane on which it exists to-day. The time is at hand when a broader viewpoint should be taken. The foundation work of creating the breeds has been done and done well, but there is still a great deal lacking."[1]

A more general use of crossbreeding would mean a larger demand for purebred stock. Little is to be gained by crossing animals that are not closely selected and uniform in their breeding behavior. Furthermore, registered livestock, when used for this purpose, can be more closely inbred than it now is. Some loss in size and vigor will occur, but the resulting crossbreds will in most cases be improved in uniformity and will lack nothing in ability to grow. Less

[1] JONES, DONALD FORSHA, "Genetics in Plant and Animal Improvement," pp. 521–522, John Wiley & Sons, Inc., New York, 1925.

attention needs to be paid to the fine points that determine the appearance of the animal and therefore its sale. Pure-bred animals that make a low score in the show ring often perform as well or better than the prize winners when used for crossbreeding.

At the present stage of our knowledge it is impossible to state with any high degree of accuracy the comparative value of the different strains of the major breeds when they are used for crossing. This can be determined only by trial, and it is important that the animals be tested and kept up to a definite standard of performance. Once a good combination is found, there is evidence to show that it will continue to give equally satisfactory results.

Finally, as Wentworth says: "Development of a system of crossbreeding would create a wider demand for purebred animals than we have at present, would eliminate much of the cost of development of purebreds since relatively few would ever have to pass good breeding condition, and would make improvement in livestock type more than ever the work of specialists. The general adoption of a system of crossbreeding would result in a better quality of breeding animals on the general farm, a better quality of commercial animals, cheaper production costs, and a more accurate reflection of market values in purebred values."[1]

In many parts of the world meat is a luxury and will probably become increasingly so. Since it takes several pounds of plant material to make a pound of meat, countries that are densely populated rely principally upon food from plants. The cereal grains combine nutritive value with high production and convenient storage. In Asia, rice has been the mainstay of the most people, just as in America maize, in Africa millo maize, and in Europe wheat, have been the leading crops. Wheat is the outstanding

---

[1] *Record of Proceedings, American Society of Animal Production, Annual Meetings,* 1925–1926, p. 47.

cultivated plant of the world at the present time. It produces the largest total amount of food and lends itself so easily to the use of machinery that it is grown cheaply on large areas.

A history of the development and use of this premier cereal would tell much about the progress of the white race. The nearest relative of wheat grows wild in Palestine. There the climate is comparatively mild. It is somewhat of a mystery how this plant has been made to endure the cold of Russian steppes, the dampness of English fogs, and the scorching of Kansas droughts.

One of the notable figures at the Kansas State Agricultural College, where I was a student, was Professor Herbert Roberts. He was often seen with a large leather bag, hung from a strap over his shoulder, full of small bottles of wheat. These wheats he had collected from all parts of the world—from Turkey and Russia principally, because from these countries had come the varieties best adapted to Kansas conditions and those that made the finest bread. These samples of wheat were planted in small plots, hundreds of them. When they grew to maturity, some were found to be uniform in type, but many were badly mixed. Following the success of Nilsson in Sweden, Vilmorin in France, and Hays in Minnesota, Roberts selected individual heads and increased the seed from them until a sufficient amount was obtained to judge their good and poor qualities when grown in comparison. Each lot was planted in a small square. The test field looked much like a gigantic checkerboard; light-colored, beardless wheats alternated with bearded wheats with dark chaff. Year after year these trial lots of wheat were grown and the poorest discarded, although it was almost a hopeless task to try to pick the best when so many yielded about the same.

In the spring of the year, when the wheat was first pushing up to head, Professor Roberts usually came late to class, his mind on those plots of wheat, trying to decide which were the best and whether any of them were really any better than the wheat the farmers were already growing.

One good wheat year followed another, but finally a winter came when the Kansas wheat crop did not look so promising. There was little snow, and the cold came early. The wind blew the soil away from the roots. In the spring the worst fears of the farmers were realized. Field after field showed only a sprinkling of green plants, not enough to pay for the harvest.

At Manhattan everything looked discouraging. Instead of the usual checkerboard, only here and there appeared anything green. The farm foreman wanted to plow up the whole field, for already the weeds were starting to grow where there was no wheat. But in one spot in the field Roberts saw a square that was all green. There could be no weeds in that patch because there were no missing plants. Looking at the label, he knew that this plot had come from a single plant, number P 762. The records showed that P 762 was a head selection from a mixed lot of wheat from Crimea. Perhaps the winters in that part of Russia were as severe as in Kansas, and maybe the winds blew just as hard. Anyway, whatever constituted winter-hardness, this wheat had it.

The field was left unplowed. The weeds grew in the rest of the field. The students scoffed: "If my father couldn't grow better crops than that, he couldn't afford to send me to this college." But the plants in the one little patch stood proudly erect as if they knew some one were watching them.

In due time several quarts of plump, red kernels were harvested. Would they make good bread? There was all too little seed for sowing, so none could be spared for baking. But perhaps another year . . . The following year was also trying for Kansas wheat, but number P 762 again lived through the winter and yielded well, and this time there was enough grain to send a small portion over to the milling and baking department. Soon the official report came back: "P 762 shows a distinctly higher protein content both in the wheat and in the flour, a higher percentage of flour than Kharkof but somewhat less than Turkey, a loaf expansion practically equal to Kharkof and slightly greater than Turkey, color of loaf equal to or greater than either of the standard varieties, and texture of loaf equal or superior to either of the other varieties."

In 1914, P 762 was given the name Kanred and distributed to farmers in the hard winter wheat sections for trial. Careful tests on the various sub-stations and on cooperative farms had shown that Kanred ripened earlier, was freer from stem rust, and markedly more winter hardy than the wheats commonly grown. In sixty-six tests Kanred yielded an average of 3.7 bushels more per acre than Turkey, the wheat most generally grown.

Ten years after it was first introduced, Kanred was grown in twenty states, on four million acres, and comprised 21 per cent of the hard, red winter wheat grown in the United States—all from a single kernel planted in 1905 in the wheat nursery of the Kansas Agricultural Experiment Station.

During those same years that the unnamed, fall-sown Kanred wheat was meeting its most severe test in winter's cold, another unknown grain was being equally tried, farther north, by shortened season and early fall frosts. To get the beginning of this story it is necessary to go back to the time of the pioneers on the plains of western Canada.

The Indians looked on in amazement while the first white settlers, sent by the Hudson's Bay Company, dug up the prairie sod with hand hoes and planted wheat and

[351]

barley, potatoes and peas. Their surprise turned to contempt when the wheat "choked with weeds and failed to ripen." In derision they called the colonists "jardiniers." For two years in succession wheat failed to make a crop, and no one would have predicted at that time that those northern plains were to become one of the most important wheat-producing areas in the world.

The failure was due, in part, to a lack of knowledge of the best methods of preparing the soil and the right time to sow. The seed did not germinate properly, and consequently the fields were foul with weeds. But the most serious handicap was the lack of varieties properly adapted to the severe conditions of summer drought and of early fall freezes. The wheat of British origin withered in the summer heat and grew too slowly to ripen before the first fall frosts, which come early in that part of Canada.

In the United States wheat sown in the fall is cut from June to August. Farmers are sometimes anxious about midsummer drought but never fear that the crop will be damaged by frost at harvest time. North of the Canadian border wheat must be sown in the spring; and the plants start to head in August and ripen in September. A crop that would make forty or fifty bushels of grain, if only it had a few more days in which to ripen, may be ruined by freezing before it is ready to cut, or snow may flatten it to the ground, leaving it hopelessly beyond the power of any harvesting machinery to gather.

The need for quick-growing, early ripening wheat has been so urgent that Canadian farmers have always been on the lookout for new varieties that would give them good yields as well as the quality of grain that the foreign markets demanded. In time, by that slow process of trial and failure, they gradually collected those wheats that ripened in most of the years and produced a profitable crop. One of the most widely grown of these early varieties was

[352]

the Red Fife wheat. Its coming to Canada illustrates the way cultivated plants are spread around the world.

David Fife, living on a farm in Ontario about 1842, obtained from a friend in Scotland a small quantity of wheat that had been taken from a ship from Danzig. The wheat had been grown somewhere in North Central Europe. The quality of the grain attracted Fife's friend. The wheat arrived in Canada in the spring and was planted at once, but it proved to be a winter wheat. It all failed to ripen except three heads that had grown from a single grain.

Seeds often get mixed in handling. They lodge in the crevices of threshing machines and are carried from farm to farm. The corners and cracks of farm wagons, warehouse bins, and the holds of ships contain many a kernel that mixes in with a succeeding shipment. In some way one kernel of spring wheat got into the seed sent to Fife and was easily detected when the ripening heads stood out in a field that was otherwise barren. The few seeds harvested were of considerable interest to Fife, although most farmers would have passed them by. Saved and sown the following year, they grew well and were entirely free from rust in a year when all the wheat in the neighborhood was badly diseased.

Although planted rather late and in an unfavorable spot, these few plants ripened a good crop. From this small beginning has come the wheat that has been grown from the Great Lakes to the Rocky Mountains as the Scotch Fife, Red Fife, or Glasgow wheat. A comparison has shown that Fife is identical with a variety grown in Europe and known as Galician. Doubtless it could tell of many interesting events in its long history from the wild.

The early distribution of this wheat was rather slow, but in the early eighties it became the principal variety in Western Canada and established the reputation of the Dominion as a producer of high quality milling and baking wheat. Manitoba No. 1 hard, the trade name for the best of this wheat, brought the highest price at Liverpool. Valuable as this wheat from Europe has been, its greatest value now lies in the early ripening Marquis wheat that has been derived from it.

While Fife was earlier than most of the wheat previously grown and freer from rust, it was still not early enough to escape damage from frost as the farms were extended northward. Summer drought withered its stalks, and in those years when rust was bad, it did not escape injury.

Before 1850 several attempts were made to establish experimental farms, but these all failed until the present Dominion Experimental Farms system was established in 1888. William Saunders was the organizer and first director. Before taking charge, he had been a druggist with a hobby for developing new fruits by crossbreeding. His grapes, gooseberries, currants, and raspberries were hardier and yielded more in the Canadian farm gardens than the varieties from the Old Country or from the United States. But what the farmers on the newly plowed prairies wanted most was an earlier maturing wheat. The soil was wonderfully fertile, and the wide, level fields were admirably suited for wheat growing on a large scale, if only the menace of the frost and snow could be averted.

Saunders collected wheats from many different countries and tested them alongside the standard Fife on the newly established experimental farms. Wheat from the high plateaus of India ripened much earlier, but the plants were small and the yield too light to be of much value. One wheat, in particular, from the Lake Ladoga region north of Petrograd, was especially promising for its earliness; but when a sufficient increase was obtained to make a milling and baking test, it was found to be seriously deficient in color and texture. Thus the expectation of replacing Fife with earlier ripening wheats from other countries came to nought.

But Saunders was not discouraged. He knew how to combine size from one fruit with hardiness from another by crossbreeding and selection. If it worked with gooseberries, why wouldn't it work with wheat? The first crosses were made at the Central Experimental Farm at Ottawa in 1888, using the imported plants from India and Russia with the locally grown Fife. Busy with administrative details, he employed his two sons, Charles and Arthur, to help him, along with J. L. McMurray and W. T. Macoun. Cross-pollinations were made at Brandon and Indian Head in Saskatchewan and at Agassiz in British Columbia. The crossed seeds or their progeny were later assembled at Ottawa, where the task of selecting and comparing the new plants was begun.

William Saunders knew how to work with vegetatively propagated fruits. When he found a plant that he liked, all he had to do was to root cuttings, and the new variety, when tested, was ready to distribute. Each plant remained true to type. The seed-propagated wheat was a different problem, as he soon found out. The generations following the cross were bewildering. No two plants were alike. An early selection of one year ripened late the next. Little dwarfs grew mixed with tall stems, and red and white kernels came from seed all selected for one color.

Before 1900 there was no exact knowledge about the behavior of hybrids. The best procedure to get them fixed in type, so that they would produce a uniform crop year after year, was not understood. But soon after 1900, Mendel's experiments became known. Nilsson, in Sweden, had shown the best way to purify a mixed variety of wheat and other self-fertilized plants by the progeny performance record.

In 1903 Charles Saunders, son of William, was placed in charge of the cereal investigations in Canada. Trained in the methods of science at Johns Hopkins, he went at the wheat breeding program from a new viewpoint and with a fresh enthusiasm. He carefully re-selected the various lots of mixed wheat, descendants of those crosses he had helped to make several years before. In order to make the plants uniform, he took single heads and planted the seeds in a separate plot, following the method of Hays in Minnesota and Nilsson in Sweden. If the offspring were again variable, but too promising to discard, another single head was selected. Realizing the importance of knowing something about the bread-making qualities of these new wheats as soon as possible, he chewed the kernels—an old trick learned from practical wheat farmers. If the resulting gum was light in color and rubbery, flour from that wheat could be expected to make good bread. Elasticity in flour paste is an indication of high gluten content, the gluten being a mixture of complex proteins that are present in considerable amount in the kernels along with the starch in those varieties that make the best flour.

Saunders discarded those wheats that chewed into a dark paste or did not make a good gum and all those that did not ripen earlier than Fife. Stiffness of straw, color of kernel, and freedom from rust were all carefully noted. Those that were lacking in any respect were left out at planting time. But still, out of some 700 lots of crossbred wheat, many remained. Growing in the experimental plots at Ottawa, they appeared to be equally good.

The one wheat from this collection that was later to be grown by the millions of bushels from Ontario to British Columbia, and south to the Missouri River, did not stand out as exceptional. Going back over the records this one wheat was later found to be below the average in yield for many years. At Ottawa it ripened only a day or two ahead of Fife—not enough difference to cause it to replace the old standby.

It was out on the prairies of Saskatchewan that this conquering cereal was to meet its test and come through with its golden stalks standing proudly erect. It took that terrible year of 1907, when even the earliest varieties failed to ripen, to prove its real worth to the spring wheat section of North America. It is largely owing to Angus Mackay that this wheat is known at all.

Mackay was a hard-headed settler at Indian Head in Saskatchewan who, largely by accident, had learned the value of summer fallow. It came about in this way. Back in 1885 he had finished plowing his land for wheat, but had it only half sown when a threatened Indian uprising brought the soldiers past his farm. They commandeered his horses and left him nothing to do but to watch the wheat sprout and the fields turn green on half of his farm. The other half remained black and barren save for the weeds starting here and there. Hating the sight of weeds, Mackay finally obtained one horse, too feeble perhaps for the soldiers; it was then too late to sow wheat, but he could at least keep those weeds from growing to seed.

Next year, with his horses back, all his fields were planted—the stubble land where a good harvest had been reaped the year before, as well as the land that had stood bare from planting time to planting time. This year, 1886, was to be put down as the worst the Canadian plains had yet experienced. There had been many dry years in Canada but none that blasted the stalks and shrivelled the grain as this one did. The wheat that the farmers had counted on to send back to the Old Country to compete with the best from Australia and Argentina was fit only for chicken feed—where there was any grain worth harvesting at all.

On Mackay's farm the stage was set for an outstanding demonstration, with results so clear cut and decisive that even the dullest could understand. All the wheat planted on the stubble ground was as poor as that on all the other fields stretching away on every side in one grim desolation, but the wheat on the fallowed field was green and thrifty.

[357]

Thirty-five bushels to the acre of plump sound grain were harvested.

Such good news spread far. William Saunders, looking for a man to take charge of a test farm in that district, knew the right kind of superintendent as soon as he saw him. Year after year Mackay was kept busy at the Indian Head farm, growing all kinds of crops, testing many new trees and shrubs, experimenting with new ways of growing old crops and trying new crops in the way that had been found to be the best.

While he was busy doing these things, the years sped along until 1907. By this time Charles Saunders, at Ottawa, had tested and retested those hybrid wheats until only a few of the best remained. The flour of all of these, each ground separately in a small mill, made good bread. The plants were stiff strawed and ripened early; the grain did not shatter when the heads were ripe. But before they could go out to the farmers they must be tried at the western experimental stations.

So, early in 1907, Mackay received several sacks, each holding about a half bushel of seed and each labelled with a different name. By this time planting new wheats was an old story to Mackay. While still hopeful, he had been disappointed too many times to be overly enthusiastic. Again the stage was set for a demonstration like that of the summer fallow that had proved to be so useful many years before.

Spring opened unusually late. In April, when the planting should have begun, the ground was hard with frost. All over Canada the seed wheat was weeks late going into the ground, and at the Indian Head farm, with many other things pressing to be planted, these small lots of wheat were not seeded until a month after the usual planting time. Even with this late start, all might have been well if July and August had been bright and warm, but when

[358]

the plants should have been growing the fastest, the days were cool and cloudy and the nights cold with threat of frost. On August first the thermometer went to 35°. On the night of September twelfth there was a hard freeze. All over Saskatchewan and the neighboring provinces, Red Fife, the farmer's principal crop, had just begun to turn from green to yellow. In the kernels there was too much moisture to escape the ruin wrought by icy crystals.

In the trial plots there was equal desolation—in all except one plot labelled Marquis. Looking at those golden yellow stalks standing erect, the kernels hard and plump, Mackay saw the wheat he had long been looking for.

Equally good reports from the other stations were sent back to Ottawa. "Marquis ripens from four to six days earlier than Fife and is less subject to stem rust." Ten years later 80 per cent of the wheat grown in the prairie provinces of Canada was Marquis. This was just in time, too, because in 1917 the mother country needed food as she never had before. After the war, a varietal census in the United States showed Marquis to be the leading spring wheat. It now makes up two-thirds of all the grain of this type harvested.

Such is the way new wheats are made from old.

Although wheat is of more immediate concern to the people of the United States, corn is indirectly of greater importance. Cattle, sheep, and swine are fattened on corn. In the form of meal and silage it makes up the bulk of the dairy ration. Without this cereal, eggs would cost more than they do. Cornstarch, corn oil, and corn syrup are used by cooks in daily recipes. Corn is the most valuable of our food resources.

A few years ago the editor of *Wallace's Farmer* reproduced on the cover of his magazine a photograph of the prize-winning ears of dent corn exhibited at the International Grain Show at Chicago. These splendid specimens—long,

[359]

broad, cylindrical ears, well capped at base and tip, the kernels deep, compact, and in straight rows—had been grown in Indiana, and represented the best from the whole country. Along with these handsome ears was printed a picture of the same number of other ears that had been grown in Iowa. No two ears were alike except that they were all poor. Of all the miserable, nubbiny, mouse-eaten, degenerate ears of corn that had ever been grown, these were certainly the worst—irregular rows, crooked cobs, bare tips, yellow, white, and red all mixed together. They were ears that most Iowa farmers would be ashamed to husk. Yet this editor of a farm paper had the effrontery, not only to show them to the readers outside of Iowa, but even to offer to let some one plant them for seed in comparison with the grand champion ears pictured alongside. Any one could see which looked the best. He wanted to know which yielded the most.

Keen-eyed Wallace had been telling the farmers of Iowa for a number of years that "looks mean nothing to a hog," but this was going too far. Hadn't they been taught that nothing but the very best ears should be saved for seed? Perhaps this young editor of an old established farm paper had lost his head entirely.

Whatever the grower of the prize-winning ears may have thought, he at least did not accept the offer to compare his good seed on a yield basis. Either, in his mind, the contest wasn't worth bothering with, or possibly there was more behind those scrawny ears than could be told by looking at them. There was. Confidentially, they were no culls from the corn crib. They were selected purposely to look worse than usual, but at best they were nothing that a corn grower would ever think of planting for seed unless he knew something of the story of crossed corn.

This story began about fifteen years before, when investigators at the Carnegie Institution on Long Island and the

Connecticut Agricultural Experiment Station at New Haven put paper bags on the ear shoots of corn before the silks appeared in order to prevent natural pollination. Bags were also put on the tassels to collect pollen. At the right time the plants' own fertilizing particles were dusted on the protected flowers. This operation, repeated year after year, resulted in the closest kind of inbreeding.

Most farmers know that inbreeding is injurious both to plants and animals; they could have told these scientists that they were wasting their time—in fact, worse. They were right; production went steadily down. After a few generations of this self-fertilization, many plants were so weak that they died before reaching the tasseling stage. Those that did live were small and feeble. Instead of the large, fat ears that corn should have, only nubbins were harvested.

But these men were interested in something more than yield. They wanted to know why the yields went down, how far they would be reduced, and in what way the inbred plants would differ from naturally pollinated plants. Then, too, what would happen when these inbred plants were crossed with each other? They tried crossing and were themselves astonished at the vigorous growth and increased productiveness of the intercrossed plants the very next year after crossing. They noticed, also, that the crossed plants of any one combination of inbred parents were closely alike; each plant grew to the same height, silks and tassels appeared at the same time, and every plant, when rightly grown, had good ears on every stalk. A row of this hybrid corn looked like a line of West Point cadets on parade. The uniformity was an indication of the high production from every plant, a very important factor in the extraordinary high yields that have since been obtained from this crossed seed.

This was the kind of seed that Wallace offered to compare in productive value with the best that the corn farmers

[361]

had been able to produce. His inbred strains were collected from several places; some he had produced himself. They had all been tested in various combinations. The nubbins that he pictured on the front page of his magazine had been grown on inbred plants and were consequently poor, but in each kernel there was a hybrid union of different heredities, purified by inbreeding. Each by itself was weak, but in combination, powerful.

In every district in Iowa—the United States Department of Agriculture has made careful yield tests—this crossed corn has given more bushels of good sound grain than the best of the varieties previously grown. Similar reports are coming from other states. Canners and market gardeners are finding that the same method applied to sweet corn gives astonishing results. The evenness in size and shape of ear and the ability to ripen at the same time are of even greater importance for sweet corn than for field corn.

Seed growers are taking an active interest in this new method, and fields used for the production of crossed corn seed are being planted with the two inbred types to be crossed in alternating rows. In midsummer, as soon as the tassels appear and before any pollen is shed, crews of tassel pullers go into the field and jerk out the opening panicle at the top of the plant from those rows that are to supply the seed. The ears on these detasseled plants receive their pollen from the adjacent rows, planted solely for the purpose of supplying pollen. This crossed seed is used only once. Later generations fall off in vigor and yield. The extra vigor and production in the first hybrid generation more than pay for the additional expense of producing seed in this new way.

In all the years that he has been growing plants and feeding farm animals, the farmer, faced with the necessity of making a living from the soil, has never carried his few

[362]

random observations on inbreeding beyond the point where production was first affected. By going farther, patient investigation, supported by the methods of science, has discovered a new way to better yields. The corn grower has known all along that the best ears saved for seed would invariably produce many nubbins. Science now shows how a bumper crop of all good ears can be grown from nubbins. But they must be the right kind of nubbins.

Chapter XII

DIET AND NUTRITION

by Elmer V. McCollum

WHAT we know about the nutritive needs of the human body and about quality in foods is the result of experimental studies on animals, correlated with such observations as can be made on human subjects. Between 1840 and about 1905, an adequate diet was generally described as consisting of appropriate proportions of proteins, carbohydrates, fats, mineral salts, and water. Oxygen of the air should have been regarded as a nutrient principle, since without it food could not be utilized; but it was not ordinarily included in the list, except as it formed part of the molecules of the foods ingested. Since this date, our knowledge of dietary requirements has advanced materially, and is still advancing, though perhaps nearing completion. The most profitable method of study has been found in efforts to simplify the diet, in order to learn what chemical substances are indispensable, and which of the compounds known to occur in the tissues are capable of being synthesized by the body.

Proteins consist of giant molecules which are resolved during digestion into about twenty relatively simple organic compounds known as amino-acids. We know that two of these can be synthesized by the body, and, pending further investigations, we may assume that the remaining eighteen must be supplied by the food. We must have the sugar glucose, which may be taken as such or derived from cane or milk sugar or from the various starches. These are all changed by the organs of digestion so as to introduce only glucose into the blood. We require at least ten inorganic or mineral elements: sodium, potassium, calcium, magnesium, chlorine, iodine, phosphorus, sulphur, iron, and copper. Probably this list of indispensable inorganic elements will eventually be extended to include several others, such as manganese, zinc, silicon, fluorine, and possibly nickel and cobalt, boron, etc. These last appear to be necessary for the development of plants, and may also be necessary for that of animals. At least six nutrient principles, called vitamins A, B, C, D, E, and G, are known to be required for normal nutrition. Among the twenty or more fatty acids known to chemists, at least twelve occur in the fats of our ordinary foods, but it appears from the results of experiment that only one of these, namely, linoleic acid, cannot be synthesized by the body from carbohydrate molecules. From the data available we may say with some assurance that the simplest diet which would furnish everything necessary for normal nutrition must contain appropriate proportions of at least thirty-six simple chemical substances. Actually, we eat daily many times this number in our ordinary foods of animal and vegetable origin.

Methods have been developed by means of which experiments on animals fed a single natural food (wheat, maize, etc.), which does not alone support satisfactory nutrition, supplemented with single or multiple additions of the indispensable nutrients, yield information concerning the nature of their deficiencies. Such experiments may be made nearly quantitative. Through such studies we have secured a considerable body of knowledge concerning the extent to which each of our more important natural foods—cereal grains, tubers, fruits, roots, leaves, meats, milk, eggs, etc.—furnishes the body with the indispensable nutrients. In addition, such experiments have shown us the nature and extent of the deficiencies of the different natural foods. On the basis of such data it is possible to theorize concerning which natural foods, individually lacking or deficient in one or another food principle, should, when combined, supplement each other's deficiencies. Experimental verification of predicted results confirms in a highly satisfactory manner the belief that we understand, to a great extent, the nutritive needs of the body and the dietary properties of most of our natural and manufactured foods.

Since most foods have been shown to be deficient or lacking in one or more of the nutrient principles, it is necessary for us to combine foods of unlike composition so that one will provide what the other lacks. Herein lies the cause of safety in variety in eating. The whole wheat kernel, which many people have long believed to be a complete food, is incapable, when fed as the sole source of nutriment, of supporting growth of the young or prolonged health in the adult. It is deficient in three respects: (1) it lacks sufficient calcium; (2) it lacks sufficient vitamin A; and (3) its proteins require supplementing with proteins from other sources which supply in abundance certain amino-acids which the wheat proteins contain in amounts too small to serve as building stones when food proteins are converted into body proteins. Even though wheat is supplemented with one or two of these substances in which it is deficient, the nutrition of an animal will not be so good as it will if all three are added to the diet.

The Vitamins. There are three vitamins— A, D, and E— which dissolve readily in fats and are found only in certain fats which our foods contain. They do not, in general, occur together. Vitamin A is abundant in cod liver oil, butter fat, and milk fat; in the glandular organs of animals, such as the liver, kidney, sweetbread, etc.; and in all yellow pigmented vegetables. It never occurs in white vegetables, such as the potato, white turnip, apple, etc.

VITAMIN A. This substance is now identified with the yellow pigment of vegetable foods, a compound known as carotin, from its abundance in carrots; or at least it seems to be demonstrated that carotin is the mother substance of the vitamin into which it is readily converted in the body. The latter seems the more probable, since the liver of an animal may be nearly freed from vitamin A by feeding a diet free from it, and become rich in the vitamin when an abundance of carotin is provided. Such a liver is still nearly free from yellow pigment. This is interpreted as meaning that the yellow pigment is converted into the vitamin, and is not itself the vitamin A. It has recently been stated, on experimental evidence, that plants are all practically free from the vitamin A, but that they furnish carotin from which it is made in the body. Liver fats, egg yolk fats, and cod liver oil contain the vitamin instead of carotin. Little is known about the chemical nature of carotin and less about that of vitamin A. The former is a highly unsaturated hydrocarbon containing 40 carbon and 56 hydrogen atoms in its molecule. It is an unsaturated molecule and takes on oxygen readily, losing in the process its yellow color and its value as the mother substance of the vitamin.

EFFECTS OF DEFICIENCY OF VITAMIN A. Much research has been done on the effects of deprivation of animals of this vitamin. The injury to the body which results from this kind of specific starvation is limited to the epithelial tissues. Since these line the ducts of the tear glands, salivary glands, and other digestive glands, and constitute other glandular tissues of major importance, vitamin A deficiency quickly undermines health. The epithelial cells keratinize, becoming like the outer layers of the skin, and lose their normal functions. Plaques of these cells desquamate and tend to plug the ducts of glands. In vitamin A

[367]

deficiency the earliest symptoms are deficiency of tears, dryness of the eyes, and dryness of the mouth. The skin becomes dry and scaly, the germinal epithelium in the testes degenerates, and the animals become sterile.

Attention has repeatedly been called to the occurrence of large numbers of calculi in the kidneys and bladders of rats suffering from deficiency of vitamin A. Rats develop deposits of phosphates and oxalates in the urinary tract very rapidly, and almost invariably when fed diets deficient in this vitamin. When chickens are fed an A-deficient diet, there accumulate in the kidneys great numbers of crystals of urates, or salts of uric acid, so that the kidneys feel sandy between the fingers. There is also some evidence that gall stones are more likely to develop under conditions of vitamin A insufficiency than otherwise. Plaques of epithelial cells desquamate and form nuclei upon which cholesterol deposits.

One of the earliest observed effects of vitamin A deficiency was the appearance of an ophthalmia characterized by drying of the cornea followed by ulceration and perforation of the eyeball. This has been shown to be a secondary result of injury to the tear glands. The gland atrophies and loses its power to secrete tears; the eyeball thereupon becomes dry, and cornification of the cornea soon develops.

A number of papers have been published which refer to the incidence of a similar ophthalmia in human subjects subsisting upon diets of poor quality. There is much reason to believe that the occurrence of night blindness is sometimes attributable to chronic deficiency of vitamin A. In the intestinal tract there may be impaired absorption due to injury to the epithelial cells of the wall.

In rats, when the diet is impoverished in vitamin A, the vaginal mucosa forms cornified epithelial cells continuously. Under normal nutrition there is a similar cornification limited to a brief period during which there is growth, maturation, and rupture of Graafian follicles, after which it disappears; but in A-deficient rats the desquamation of cornified cells is continuous and obscures all ovarian cycles that may be present. Vitamin A deficiency does not create a malfunction of the ovaries, for they continue to secrete hormones similar to those in estrus. This effect on the vaginal mucosa of the rat is so pronounced that Evans and Bishop report their ability to detect vitamin A deficiency in otherwise apparently healthy animals receiving enough A to prevent ophthalmia. Raising the level of A in the diet may abolish the persistent cornification of the vaginal epithelium and restore normal conditions.

In vitamin A deficiency this transformation of epithelium into stratified, squamous keratinizing epithelium is especially pronounced in the upper respiratory tract, and in the renal pelvis, urinary bladder, seminal vesicles, epididymis, prostate, salivary glands, and pancreas.

When there is deficiency of vitamin A, the intestinal flora is markedly changed as respects gram-negative and gram-positive bacteria in the faeces. The faeces of rats on the deficient diet are dry and hard, which may account for the disappearance of streptococci; otherwise there is no change in the proportion of bacteria which ferment glucose, lactose, and sucrose. The proportion of hydrogen-sulphide-forming bacteria remains constant.

Under deficiency of vitamin A, rats frequently die of bacterial invasion of the ear and nasal cavities before the appearance of ophthalmia. As the infection advances, it leads to nutritional disaster in which the animal is not restored to a normal condition by feeding rations containing vitamin A. Werkman has reported that a deficiency of vitamin A in the diet increased the susceptibility of rats to anthrax and pneumonia.

An interesting observation reported by Howe is the reversion of the odontoblasts to osteoblasts when animals

[369]

are deprived of vitamin A. The odontoblastic membrane surrounds the pulp of the tooth and lies in apposition to the under surface of the dentine. From each odontoblast a fiber extends through a tubule in the dentine to the base of the enamel. The odontoblast forms dentine. Howe states that when animals are deprived of vitamin A, the odontoblasts revert to osteoblasts, or bone-forming cells, and that subsequently bone deposits may be formed by these cells. The frequent occurrence of pulp stones in the teeth of animals deprived of vitamin A supports this view. The importance of this vitamin for the health of the teeth appears, therefore, to be very great.

VITAMIN B. Our knowledge of the specific effects of deficiency of vitamin B, which is one of several water-soluble nutrients, is less satisfactory than in the case of vitamin A, because uncomplicated vitamin B deficiency has less frequently been produced experimentally. By vitamin B we designate the anti-beriberi principle, or the anti-neuritic vitamin. There is little reason to doubt that deficiency of this vitamin is the cause of the disease beri-beri, which has taken millions of lives during the last 2000 years. The disease occurs mainly among rice-eating people who take polished rice as their principal food. Most of the vitamin is concentrated in the germ and the bran layer, both of which are removed in polishing. Beri-beri is still one of the chief causes of infant mortality in the Philippine Islands and in the East Indies.

In vitamin B deficiency the nervous tissues suffer an injury almost as specific as that suffered by the epithelial cells in vitamin A starvation. The peripheral nerves are first affected. They undergo degeneration, and the muscles innervated by these nerves undergo atrophy. For these reasons beri-beri patients suffer paralysis.

An interesting effect of B deficiency is lack of appetite. Experimental animals cease to eat their caloric require-

[370]

ments, but the appetite improves immediately on administering the extract containing the vitamin. All species of animals appear to require this vitamin.

Vitamin B is abundant in yeast, in germs of cereals, and in the leaves of young plants. It occurs generally in tubers, roots, fruits, leaves, cereal grains, milk, eggs, and the glandular organs of animals. Muscle meats are almost lacking in it; and refined cereal products, such as white flour, degerminated corn meal and polished rice, the sugars, starches, and fats from both animal and vegetable sources are devoid of it. Recent studies by Macy, in which careful assays of both human and cow milks were made by feeding tests with young rats, have shown that about seven times the volume of milk is necessary to provide the minimum of vitamin B on which approximately normal growth can take place than is required to furnish an adequate amount of vitamin A. Her studies suggest that infants restricted solely to a milk diet during the first six months of life are developing on the minimum intake of vitamin B on which growth would be possible. It appears that infants would probably fare better if their milk were fortified with some concentrate containing this vitamin.

VITAMIN C. Vitamin C is the anti-scorbutic principle. This is the most unstable of the vitamins and is especially susceptible to destruction by contact with the oxygen of the air. Birds apparently do not require this vitamin, and this has also been demonstrated for the rat. In these species the body has the capacity to synthesize the vitamin from some other component of the diet, since the livers of rats grown to maturity on diets free from C are very potent in curing acute scurvy in guinea pigs.

Deficiency of vitamin C produces characteristic pathological changes. It appears that the endothelial cells lose their power to produce the cement substance which holds them together; and so the capillary blood vessels, whose walls consist of a single layer of endothelial cells, break down, allowing the blood to ooze out into the tissues. Hemorrhage is therefore the most outstanding change resulting from vitamin C deficiency; but another effect, namely, rarifaction of the bones, is of great importance. While scurvy is developing, the bones become rarified and fragile. Rarifaction of the alveolar bone which forms the sockets of the teeth is the cause of looseness of the teeth seen in scurvy.

Approximately forty days of total deprivation of vitamin C causes the appearance of the clinical signs of scurvy. Höjer made the interesting observation that in approximately half the time required to produce clinical scurvy the odontoblastic layer in the teeth is affected. If approximately seven-tenths or eight-tenths of the minimum protective dose of vitamin C is given to guinea pigs for ten to fourteen days, the odontoblastic membrane will show abnormal structure, the cells being of unequal size. If but half the minimum protective dose is given, the odontoblastic layer will have separated from the dentine, the fibrils apparently having become detached. If but three-tenths, or thereabouts, of the protective dose is given, the odontoblastic layer will have broken up into islets of distorted cells. This may be of great significance to the health of human teeth, since doubtless many people, at different times, have gone for more than twenty days entirely without a supply of vitamin C. If changes comparable to those in the guinea pig take place in the human teeth, it is not unreasonable to suppose that subsequent death of the pulp would be the result of this type of specific starvation. The frequent occurrence of lifeless pulps in otherwise normal-appearing teeth demands an explanation as to its etiology. The recent researches on animals showing odontoblastic damage seem to account for the condition seen in many human teeth.

Vitamin C is most abundant in fresh vegetable foods, but the livers of well-fed animals contain a store of it. Pasteurization of milk destroys practically all of it; and as pasteurization of milk has increased in cities, infantile scurvy has likewise increased. This is not a sound argument against pasteurization, since it is now a general practice to give infants some suitable fruit or vegetable juice which will supply what has been destroyed in the heat treatment of the milk.

Citrous fruits and tomatoes are outstanding as rich sources of this vitamin. Ordinary canned tomatoes are little inferior to the fresh article, but bottled sterilized orange and lemon juice have generally been found to be without value in this respect. Raw potatoes, raw cabbage, lettuce, turnips, etc., are excellent sources of the anti-scorbutic vitamin.

The readiness with which vitamin C is destroyed on heating milk and the absence of the vitamin from foods cooked in the ordinary way led, a few years ago, to the unwarranted assumption that canned fruits and vegetables would be worthless as anti-scorbutic foods. That this was an error was pointed out by Kohman and Eddy, who showed that modern canning processes tend to preserve much of the vitamin C content of the foods. The destruction in ordinary cooking results from heating the food while it still contains, dissolved in its juices, about 5 per cent by volume of oxygen, and from continued heating in contact with air. If this oxygen is removed by immersing the food for a few hours in water before processing, the subsequent heat treatment does not destroy the vitamin. Modern canning methods involve subjecting the foods to diminished pressure and to removal of the air by steam. This system of cooking is responsible for the preservation of the anti-scorbutic quality of most canned foods.

[373]

VITAMIN D. Previous to 1922 nearly all infants and young children in the north temperate zone suffered at least a mild attack of rickets. In this condition the bones do not calcify and, as a consequence, are subjected to deformity. The cause of the disease was long a mystery, although it had been recognized for more than one hundred years that cod liver oil was a specific for its prevention or cure. It was not until rickets had been produced experimentally in animals that much progress was made in understanding human rickets.

Rickets develops in young animals when they are fed a diet deficient in calcium, excessive in phosphorus, and deficient in vitamin D—a substance which is abundant in cod liver oil. Rickets also develops when the diet contains little phosphorus, an excessive amount of calcium, and a deficiency of vitamin D. In this disease the phosphorus content of the blood is markedly reduced, the inorganic phosphate being as low as one-fourth the normal eight milligram per hundred cubic centimeters of plasma. Feeding cod liver oil tends to raise the phosphorus content of the blood to the normal level.

When the diet is deficient in calcium and contains excessive phosphorus, the phosphorus is absorbed into the blood but cannot be deposited in the bones because there is no calcium to combine with it. This phosphorus is excreted as phosphate to a considerable extent into the intestine and carries with it some calcium irrespective of the needs of the body for this element. Rickets so produced tends to be complicated by tetany, a condition in which the blood calcium is reduced. In the reverse of this condition—that is, when a low phosphorus, high calcium diet is fed—an excess of calcium is absorbed in the blood but cannot be retained and is excreted into the intestine, taking with it a certain amount of phosphorus irrespective of the needs of the body for this element. This quickly reduces the content of phosphorus in the blood. The provision of vitamin D tends to regulate the calcium and phosphorus metabolism so as to keep the blood at a normal concentration as respects these elements.

While vitamin D occurs in abundance only in cod liver oil, and to a lesser extent in egg yolk, the mother substance of it, ergosterol, occurs widely distributed in plant products. It is most abundant in the fungi and was originally discovered in the oil of ergot. Ordinary yeast contains about 0.15 of 1 per cent on a dry basis, but may be cultivated under conditions where much higher yields are obtained. Ergosterol is chemically a sterol related to cholesterol, which is a component of every living cell and is especially abundant in nerves and brain. Ergosterol is a white crystalline substance which, when administered to a rachitic infant or rat, exerts no beneficial effects if the substance is kept in subdued light. If, however, the ergosterol has been exposed for a brief period to a source of ultra-violet rays, it is activated or changed into vitamin D. Only wave lengths between 265 and 313 microns are effective in activating ergosterol. This discovery constituted the first demonstration that a vitamin is a chemical substance. Activated ergosterol is now prepared in large amounts and is marketed dissolved in some vegetable oil which is palatable and does not easily grow rancid. The commercial product is known as Viosterol and is, to a considerable extent, now being prescribed by physicians instead of cod liver oil. It is recognized, however, that the diet of the child must provide a liberal amount of vitamin A in order to make Viosterol therapy equivalent to cod liver oil therapy, since cod liver oil is rich in both A and D.

The discovery of ergosterol and its activation by ultra-violet rays explains the relation of climate to rickets. Rickets has been rare in the Tropics because there people expose a considerable area of their skin to a sunlight rich in ultra-violet rays. These rays are sufficiently penetrating to activate the ergosterol in the skin and so to protect the development of the bones. Peoples in the Far North, who also escape rickets, do so because they take vitamin D in the oils of marine origin, eggs, etc.

VITAMIN E. There is widely distributed in vegetable products a substance which has been shown to affect reproduction in the rat. Female rats deprived of this substance exhibit normal oestrus, ovulation, copulation, and implantation. In gestation where E is insufficient or absent, the embryos at first appear normal but show retardation in development by the eighth day or soon thereafter. There is no detectable abnormality of structure, but sometime between the twelfth and twentieth day foetal death occurs, usually on the twelfth or thirteenth. The maternal placenta is not altered structurally sufficiently to indicate that its function is impaired. In males kept on an E-free diet, sterility develops after about the seventieth day. Spermatogenesis ceases and spermatozoa are replaced by spermatids. Sterility in the male once developed is permanent. The testes of such rats undergo a slow progressive desquamation and degeneration of the germinal epithelium involving chromolysis, plasmolysis, pycnosis, giant-cell formation, karyolysis and karyorrhexis of the germ cells followed by lipoid degeneration, granular liquefaction, and final dissolution of the cells.

Vitamin E is soluble in fats and is most abundant in the oil of wheat germ. Apparently all cereals and leafy vegetables, and probably root vegetables as well, contain a considerable amount of this principle.

VITAMIN G. Pellagra was not recognized in the United States until 1908. From that time it increased rapidly, especially in the southern states. Numerous efforts have been made to discover the etiology of the disease, which is characterized by stomatitis, diarrhea, dermatitis, and

[376]

thickening and bronzing of the skin where exposed to light. Nervous symptoms also become prominent as the disease develops. A characteristic of pellagra is its tendency to develop in the spring, and it is much more common among people in low economic circumstances than among the well-to-do. The outstanding research on the subject was carried out by Goldberger between the years 1922 and 1926. He seems to have established that it is due to deficiency of one of the water-soluble vitamins, which is now designated by the letter G. Some British workers speak of it as the anti-dermatitis vitamin. It is much more stable toward high temperatures than is the anti-beri-beri vitamin B and survives heating under pressure at a temperature of about 118° for three or four hours. Goldberger found yeast to be the most potent source of vitamin G. He believes yeast, eggs, meat, and milk, in the order given, to be effective foods in the prevention or cure of pellagra. Yeast administration is widespread in areas where pellagra is endemic, and appears to be effective where taken in sufficient amounts.

Iodine Deficiency and Goiter. One of the outstanding discoveries in the field of nutrition in recent years is the relation of deficiency of iodine in food and water to the incidence of simple goiter. Numerous losses of young farm animals throughout the Great Lakes region northwestward into Montana and elsewhere have been due to this cause. The soil in these goitrous regions is practically free from iodine so that plant products grown thereon do not contain it. Injury from iodine deprivation results before birth as well as after. The administration of a suitable amount of iodine to the human population in goitrous areas bids fair to correct this trouble. Kimball has emphasized the inadequacy of iodine therapy for infants born of mothers whose thyroids were deficient in iodine. He points out that the thyroid under these conditions is injured in pre-natal life and that subsequent provision of iodine will not prevent some abnormality in the gland. The importance of keeping the expectant mother's thyroid provided with iodine is, therefore, apparent.

Secondary Anemia and Iron Assimilation. Secondary anemia may be due to excessive hemorrhage, hookworm, chemical poisoning, etc. The impoverished blood tends to return promptly to normal when an adequate diet is provided. An interesting discovery made by Hart and co-workers relates to the conditions under which iron can be assimilated and converted into the respiratory pigment of the red blood corpuscles. They have shown that iron assimilation is impossible in the absence of a small quota of copper in the body. When the diet is entirely adequate in every respect, including copper, inorganic iron salts are capable of utilization for hematin synthesis. The relative values of iron salts depend only upon their solubility; although it was demonstrated some years ago that, if ferrous salts are introduced into the food, oxidation of vitamin A by the air is strongly catalyzed. For this reason animals may be fed an adequate amount of A, but if the iron is in the ferrous form, the vitamin A may be destroyed and the animals will then develop the typical ophthalmia of dietary origin.

Until recently, pernicious anemia did not yield to any form of treatment. Minot and Murphy discovered that when pernicious anemia patients are given large amounts of liver daily, the blood stream promptly improves and the anemia disappears in great measure. The disease is not curable; and liver, or a suitable preparation from it, must be taken throughout life, otherwise a return of the anemia is certain.

All pernicious anemia patients have a long history of digestive disturbances terminating in a loss of capacity of the stomach to produce hydrochloric acid. It has been shown that in the digestion of meat in the normal stomach something is formed which is indispensable to the proper functioning of the structures in the bone marrow which form red blood corpuscles. The capacity to produce this substance is lost by the debilitated stomach of the patient with pernicious anemia. It appears that this substance formed in the stomach is largely absorbed in the intestine and tends to accumulate in the liver; but a portion passes on to the general circulation and tends to accumulate in the kidneys. The kidneys are not so good as liver for the relief of pernicious anemia, but are greatly superior to the muscle cuts of meat. The fresh stomachs of pigs, when ground and allowed to self-digest for a while before being cooked, are highly effective as a source of the substance furnished by liver. Several manufacturers are now putting on the market concentrated extracts of liver which are as effective as liver itself in improving the condition of the blood of pernicious anemia patients. The nature of the substance is still unknown.

Supplementary Values in Proteins. Chemical analyses of proteins isolated from various animal and plant sources have revealed a surprising difference in their constitution. Human muscle contains about 12 per cent of a digestion product called glutamic acid. Wheat proteins yield as much as 45 per cent of this substance. This is an example of a food protein yielding an excess of one of the digestion products. There are many illustrations of food proteins yielding too little of one or more digestion products which are necessary for the formation of body proteins. The quality of the protein is determined by its content of that indispensable amino-acid which is present in the smallest amount. Because of the excess or deficiency in the yields of amino-acids, most food proteins of vegetable origin are not of very high quality—that is, these food proteins cannot be transformed with high efficiency into animal

[379]

proteins. The farm pig fed only wheat proteins can transform about 23 per cent of them into pig protein, whereas it can transform about 65 per cent of the proteins of milk into pig protein. Since proteins differ so widely in their composition, it is frequently found that combinations of two or more proteins from different sources have a higher biological value than either one fed singly. This is because they are not equally deficient in any one amino-acid. Feeding tests have shown that the proteins of the different cereal grains tend to be deficient in the same digestion products, and, accordingly, the combination of wheat and corn, of wheat and oats, or of corn and oats proteins does not enhance the value of the mixture. The protein mixture is not enhanced by combining pea or bean proteins with any cereal grains or by combining pea and bean proteins alone. On the other hand, the proteins of eggs, meat, and milk are especially rich in those digestion products which are furnished in but small amounts by most vegetable proteins. Hence, combinations of these proteins of animal origin with cereal or other vegetable proteins make mixtures of high biological value. The proteins of the leaves of plants are so constituted as to enhance markedly the proteins of cereals, peas, beans, and other seeds of plants.

There has long been much speculation concerning the optimum content of protein in the dietary. The vegetarian generally eats a relatively low protein diet; the meat eater a high protein diet. Experiments have shown that even a single large dose of histidine, tyrosine, or cystine—three digestion products common to most proteins—when injected into the blood of a dog will seriously injure the kidneys. Experiments on rats have shown that high protein dietaries, continued over a considerable period, tend to produce nephritis. Hinhede states that the occurrence of nephritis parallels the protein consumption in

Greenland, Iceland, Faroe Islands, and Denmark. These areas are named in the order of decreasing protein intake.

A standard argument against a luxus consumption of protein is based on the excessive putrefaction of undigested protein residues in the colon with the consequent absorption of a series of products of bacterial origin which have no place in physiological processes and which are regarded by some as physiological abominations. The prevailing view among experimenters appears to be that a relatively low protein dietary furnishing proteins of high biological value is more likely to promote well-being than is the dietary containing an excess of protein.

The Calcium Content of Foods. All seed products, such as cereal grains, peas, beans, all tubers and fleshy roots, all meats and eggs, are too poor in calcium to supply the physiological needs of growing young. The only foods rich in calcium are the leaves of plants and the milk of animals. All seed products, meats, and eggs contain an excess of phosphorus in proportion to calcium. This ratio, as already stated, is unfavorable to the calcification of the bones. In order to provide a diet which is appropriately constituted in respect to calcium and phosphorus, therefore, the diet should contain an appropriate amount of either a leaf vegetable or of milk.

Meats are deficient in all vitamins except G, the anti-pellagra principle. Milk contains all the vitamins, but is somewhat deficient in B. Egg yolks contain the full quota of vitamins. All foods are deficient in vitamin D; but many of them contain traces of the mother substance—ergosterol—which is converted into the vitamin, provided one is exposed sufficiently to sunlight of good quality, that is, sunlight containing some ultra-violet rays.

During the last 100 years there has been a steady increase in the consumption of sugar. This is solely an energy food, since it contains no protein, no mineral elements, and no vitamins. The consumption of sugar is now approximately 115 pounds per person per year, not including molasses, corn syrup, and honey. This is about ten times the amount consumed in the United States one hundred years ago. While sugar is not inimical to health when eaten in such large amounts, a food so deficient in necessary principles tends to crowd out of the diet significant amounts of other foods which would supply the indispensable nutrients.

Since about 1880 the practice of using a highly refined white flour has been general. This is deficient in all the vitamins, is almost lacking in calcium, contains proteins of relatively poor quality, an acid ash, and is especially rich in potassium and phosphorus. Because of improved keeping qualities, degerminated corn meal is now extensively used, and wherever rice is eaten, polished rice, which is deprived of its vitamins and much of its mineral content, is the rule. The diet of many people in America and Europe may be accurately described as one consisting largely of products made from white flour and other cereal products having similar dietary qualities, meat, potatoes, and sugar. No combination of these foods makes an adequate diet. This list and any modification of it containing roots, tubers, and fruits is supplemented in an effective way by the inclusion of suitable amounts of milk or of leafy vegetables. It is for this reason that milk and the leafy vegetables have been distinguished from other foods as the protective foods.

The sensitiveness of certain tissues to deficiency of antiscorbutic vitamin makes it a matter of great importance that the diet should contain regularly suitable amounts of uncooked vegetable food, or of food canned by a process which preserves this vitamin.

It is beyond the scope of this chapter to discuss in detail the planning of menus which will be complete so as to promote optimum well-being. The objective was to set forth the principal scientific discoveries concerning foods and the nutritive needs of the body. The application of this knowledge to the planning of the diet will be found discussed in full elsewhere.[1]

[1] McCollum, E. V., and Nina Simmonds, "Food, Nutrition, and Health," 2d ed., The Macmillan Company, New York, 1922.

INDEX

International Grain Show, 359
Intestines, 205, 206
Introversion, cause of, 114
Intuition, 3
Iodine, effects of deficiency of, 377
Ireland, 340
Irish, rating of, in group test, 145
Iron, assimilation of, 378
Italians, rating of, in group test, 145
Italy, 148
"Ivanhoe," 318

J

James, William, 96, 100, 120, 122
Japanese beetle, habits of, 306
  methods of control, 307
Jefferson, Thomas, 189
Jeffersonian democracy, 188
Jenner, Edward, 243
Jennings, 67
Judge Baker Foundation, 78
Jung, 75
Jurgen, 13
Juvenile courts, institution of, 78

K

Kansas Agricultural Experiment Station, 351
Kansas State Agricultural College, 349
Keller, Helen, 185
Kenagy, 146
Kepler, 11
Kidneys, functions of, 221, 224, 287
  structure of, 285
  treatment of, in disease, 225
Kimball, 377
Kimberly-Clark Corporation, 126
Kindergarten movement, 68
King George, 328
Koffka, K., 84
Köhler, 68, 83, 84
Kohman, 373
Kraepelin, 108
Kretchmer, 106

L

Laible, 345
Laird, 146
Lamarck, 18
Lashley, 120
Laws of Motion, 185
Le Bon, 79
Leenders & Co., 340
Leeuwenhoek, 17
Leibnitz, 58
Lenin, 126
Leprosy, 235
Levy-Brühl, 79
Liberia, 325, 327
Lime, 234
Limitation of the Linkage Groups, 172
Lincoln, Abraham, 187
Linear Order of the Genes, 172
Link, 138
Linnaeus, 18
Lippmann, Walter, 11
Liver, diseases of, 216
  functions of, 213, 214, 215
Locke, 59, 96
Loeb, 67
Longfellow, 320
Lungs, diseases of, 203
  function of, 224
Lyonet, 17

M

Machinery, power, increase in production due to, 322
Mackay, Angus, 357, 358
Macoun, W. T., 355
Macy, 371
Macy, R. H., & Co., use of psychological tests by, 136, 138
Malaria, control of, through public health measures, 242
  discovery of parasite causing, 296
  ravages of, 295
  symptoms caused by, 295

[392]

Queen Guinevere, 319
Quinine, use of, in treatment of malaria,
296

R

Rabelais, 11
Races, intermingling of, 88
struggle of, 43
Radin, 79
"Rational Evolution," 27
Read, Carveth, 38
Réaumur, 17
Recessive, definition of, 165
Reduction division, 172, 181
Reese, Alfred, 337
Reflexes, 59, 62, 63
"Religion and the Rise of Capitalism,"
50
Rheumatism, causes of, 230
treatment for, 231
Rich, 106
Rickets, causes of, 374
treatment for, 374
Riley, C. V., 305
Rivers, W. H. R., 43, 76
Roberts, Elmer, 345
Roberts, Herbert, 349
Romans, engineering undertakings of,
serve as public health works, 243
Roses, breeding of, Pernet, 319
Silver Moon, 339, 340
Van Fleet, 319, 339
Ross, Sir Ronald, 295, 296
Rubner, 220
Russell, Bertrand, 50, 292, 315
Russia, psychologists in, 148
upheaval in, 34
vast industrial experiment in, 125
wheat from, 349, 355

S

Safety, psychological aspects of, 147ff.
Salvarsan, discovery of, 235

Saunders, Arthur, 355
Saunders, Charles, 355, 358
Saunders, William, 354, 358
Scandinavians, rating of, in group
test, 145
Schleiden, 19
Schmalhausen, S. D., 89
Schwann, 19
Science, popularization of, 52
position of today, 52
warfare with theology, 52
"Science of Life, The," 52
Scientific dogmatism, 10
Scientific management, 130
Scientific methodology, 3, 5
Scientific Naturalism, 7
Scotch, rating of, in group test, 146
Scotland, 344
Scott, Walter Dill, 138, 140
Scovill Co., 139
Scurvy, causes of, 236, 372, 373
treatment for, 244, 371
Seashore, 105
Secretions of body, influence on
behavior, 21
Segard, 137
Segregation, Law of, 165
Serology, 21, 252
"Sex and Civilization," 89
Sex hormones, control of reproductive
activity through use of, 302
functions of, 301
Sex-linked characters, 182
Shellow, Mrs., 137
"Short Outline of Comparative Psy-
chology, A," 67
"Side Lights on the Evolution of
Man," 40
"Sins of Legislators, The," 50
Skill, physiological basis of, 65
Skin, diseases of, 238
Slocombe, C. S., 154
Smallpox, basis laid for control of, 243
quarantine against, 261
Smith, and Guthrie, 99